Environmental Cultures in Soviet East Europe

Environmental Cultures Series

Series Editors:
Greg Garrard, University of British Columbia, Canada
Richard Kerridge, Bath Spa University, UK

Editorial Board:
Frances Bellarsi, Université Libre de Bruxelles, Belgium
Mandy Bloomfield, Plymouth University, UK
Lily Chen, Shanghai Normal University, China
Christa Grewe-Volpp, University of Mannheim, Germany
Stephanie LeMenager, University of Oregon, USA
Timothy Morton, Rice University, USA
Pablo Mukherjee, University of Warwick, UK

Bloomsbury's *Environmental Cultures* series makes available to students and scholars at all levels the latest cutting-edge research on the diverse ways in which culture has responded to the age of environmental crisis. Publishing ambitious and innovative literary ecocriticism that crosses disciplines, national boundaries, and media, books in the series explore and test the challenges of ecocriticism to conventional forms of cultural study.

Titles available:
Bodies of Water, Astrida Neimanis
Cities and Wetlands, Rod Giblett
Civil Rights and the Environment in African-American Literature, 1895–1941, John Claborn
Climate Change Scepticism, Greg Garrard, George Handley, Axel Goodbody, Stephanie Posthumus
Climate Crisis and the 21st-Century British Novel, Astrid Bracke
Colonialism, Culture, Whales, Graham Huggan
Ecocriticism and Italy, Serenella Iovino
Fuel, Heidi C. M. Scott
Literature as Cultural Ecology, Hubert Zapf
Nerd Ecology, Anthony Lioi
The New Nature Writing, Jos Smith
The New Poetics of Climate Change, Matthew Griffiths

This Contentious Storm, Jennifer Mae Hamilton
Ecospectrality, Laura White
Teaching Environmental Writing, Isabel Galleymore
Radical Animism, Jemma Deer
Cognitive Ecopoetics, Sharon Lattig
Digital Vision and Ecological Aesthetic (1968–2018), Lisa FitzGerald

Forthcoming titles:
The Living World, Samantha Walton
Weathering Shakespeare, Evelyn O'Malley
Imagining the Plains of Latin America, Axel Pérez Trujillo Diniz
Ecocriticism and Turkey, Meliz Ergin

Environmental Cultures in Soviet East Europe

Literature, History and Memory

Anna Barcz

BLOOMSBURY ACADEMIC
LONDON • NEW YORK • OXFORD • NEW DELHI • SYDNEY

BLOOMSBURY ACADEMIC
Bloomsbury Publishing Plc
50 Bedford Square, London, WC1B 3DP, UK
1385 Broadway, New York, NY 10018, USA
29 Earlsfort Terrace, Dublin 2, Ireland

BLOOMSBURY, BLOOMSBURY ACADEMIC and the Diana logo are trademarks of
Bloomsbury Publishing Plc

First published in Great Britain 2021
This paperback edition published in 2022

Copyright © Anna Barcz, 2021

Anna Barcz has asserted her right under the Copyright, Designs and Patents Act, 1988,
to be identified as Author of this work.

For legal purposes the Acknowledgements on p. x constitute an extension
of this copyright page.

This publication has been co-financed under the program of the Polish Minister of Science
and Higher Education under the name 'National Programme for the Development of
Humanities' (2016–2019).

Cover design: Burge Agency
Cover image © Shutterstock

All rights reserved. No part of this publication may be reproduced or transmitted in any
form or by any means, electronic or mechanical, including photocopying, recording, or any
information storage or retrieval system, without prior permission in writing from the
publishers.

Bloomsbury Publishing Plc does not have any control over, or responsibility for, any
third-party websites referred to or in this book. All internet addresses given in this book
were correct at the time of going to press. The author and publisher regret any
inconvenience caused if addresses have changed or sites have ceased to exist, but can
accept no responsibility for any such changes.

A catalogue record for this book is available from the British Library.

Library of Congress Control Number: 2020038572

ISBN: HB: 978-1-3500-9835-0
PB: 978-1-3502-0064-7
ePDF: 978-1-3500-9836-7
eBook: 978-1-3500-9837-4

Series: Environmental Cultures

Typeset by RefineCatch Limited, Bungay, Suffolk

To find out more about our authors and books visit www.bloomsbury.com
and sign up for our newsletters.

Contents

List of Illustrations	ix
Acknowledgements	x
Introductory Remarks	1

Part One Unknownland: Retelling the Environmental History of Soviet Eastern Europe through Literature and Cultural Memory

1	Narrating History across Borders	9
2	History and Literature	15
3	Environmental History	19
4	Cultural and Environmental Memory	25

Part Two The Tired Village

1	Historical Background	35
2	Fatigue: Platonov's *Pit* and the Stalinocene	39
3	The Rural World is Gone: Peasants' Voices	57
4	*Satantango*: Interconnecting Human and Ecological Worlds	79

Part Three The Earth's Memory

1	Mining Narratives and Their Historical Background	93
2	Unearthing the Story of Coal: *Drach*	107
3	The Uranium Narrative: History of a Disappearance	115

Part Four The Persistence of Chernobyl in Cultural Memory

1   Eastern European Risk Narrative: Chernobyl Memorial   127

2   Contaminated Language: Wolf's *Accident*   137

3   The Bees Knew: Alexievich's Chronicle   147

Part Five Disturbed Landscapes

1   Non-sites of Memory and the Violation of Nature   161

2   Greening Sites of Memory   177

3   Białowieża Forest across Eastern Europe's Borders   189

Notes   209
Bibliography   211
Index   229

List of Illustrations

1	Andrey Platonov in 1938.	43
2	'Let's fulfil and exceed the new five-year plan! More beets mean more sugar!' Propaganda poster by Vatolina Nina Nikolaevicha. 1946.	58
3	Józef Chełmoński, *Indian Summer*, 1875.	77
4	Józef Chełmoński, *Storks*, 1900.	77
5	Endless grasslands (*Puszta*) on the Great Hungarian Plain. Author: Beroesz.	80
6	'Smoke of chimneys is the breath of Soviet Russia.' Early Soviet poster.	95
7	Halemba coal mine's shaft with Polish flag.	100
8	Eastern Urals State Reserve established in 1966 after the Kyshtym accident.	132
9	'Atom for peace.' USSR stamp: Radioactive Decay as Symbol of Atoms for Peace. Emblem and Pavilion at Expo '67.	138
10	Babiy Yar.	166
11	Białowieża Forest.	202

Acknowledgements

I would like to express my gratitude to two special institutions for their extremely important fellowship programmes of which I was beneficiary during the writing of this book. Significant thanks are due to Trinity Long Room Hub Arts and Humanities Research Institute at Trinity College Dublin. As a Marie Skłodowska-Curie COFUND Fellow for 2018–19, I was generously hosted by the Hub during the first period of preparing the book. Special thanks go to Jane Ohlmeyer for her attentiveness, and to Caitriona Curtis for her friendship. TCD Environmental Humanities team and James Smith – thank you for all your valuable thoughts.

I would like to thank Krzysztof Smólski, a managing editor of the Institute of Literary Research Publishing House in Warsaw, who administered a grant received from the Polish Ministry of Science to support part of the publication costs.

The final stage of preparing this book I shared with the vibrant and critical community of fellows from the Rachel Carson Center for Environment and Society (RCC) in Munich. I received immense support from them – special thanks go to Stephen Halsey, Thomas Lekan, Anna Pilz, Bettina Stoetzer, Anna Varga, Jessica White, Kate Wright and RCC staff.

This book would never have appeared in this form without the guidance, support and friendship of other individuals. I am indebted to the series editor, Greg Garrard, whose constructive remarks enhanced the narrative of the manuscript and who encouraged me to speak out in my own Eastern European voice. No less contribution has been made by my assistant editor, Michelle Nieman, whose enthusiastic and close reading of the final version enabled me to achieve a strong sense of the whole project. The intellectual stimulation from engaged conversations with Michael Cronin and Christof Mauch fuelled this book immeasurably. While receiving comments, advice and enormous encouragement from my mother and my closest friends, I was reminded to what extent phenomenal minds surround me: Michał Barcz, Anita Jarzyna, John Morrill, Marta Tomczok and my beloved Clever Hans – thank you! My most heartful gratitude, however, is due to my son Ignacy, whose patience and belief in me comes from his selfless love.

Introductory Remarks

The shadow of the Chernobyl disaster, villages and towns depleted after many decades of intensive coal mining, soil contamination and water pollution from collective farming and heavy industry, towns closed after radioactive uranium ore was extracted out from under them – these are but a few of the traces that the centrally planned local economies of the former Soviet bloc states left on the landscape. The Soviet Union and its satellites are rightly identified as among the twentieth century's main destroyers of the environment. During the Cold War, however, the difference between the capitalist West and the communist East in damaging nature and increasing ecological risk was a matter of scale and performance, not a result of the environmental cultures that these countries represent.

Plenty of books have been written and studies conducted on the 'anti-environmental' history of Soviet countries. I shall refer to many of them. In these publications, which are usually written by Anglo-American historians, Soviet Russia unsurprisingly dominates and provides the most often repeated examples of ecological catastrophes – Chelyabinsk, the Aral Sea and Chernobyl. It is not easy to look beyond the normative opposition of East and West as we still live in a divided world. I have written this book because I seek to retell the environmental history of this region, not from a privileged position, nor in response to new evidence, but because the environmental literature that past events prompted can help us reinterpret them.

The anti-environmental history of Soviet Eastern Europe does not adequately represent the region's human–nature relations, which I combine here with reflections on cultural memory and reinterpretations of the pastoral. Some historians who are not environmentally oriented in their studies nevertheless recognize the unique relationship between violence and the landscape in this region. Books about Eastern European history, such as *Bloodlands* by Timothy Snyder (2010) and *Kontaminierte Landschaften* ('Contaminated Landscapes') by Martin Pollack (2014), refer to mass killings committed by the regimes of Stalin and Hitler not only in the concentration and forced labour camps, but also in locations isolated by wilderness. The memory of genocide is partially – sometimes even materially – lodged in the living tissue of environments or, as in the example of the Chernobyl catastrophe, in human and ecological disasters that dramatically converged. The short history of *Homo Sovieticus* interferes with the long-term environmental history of Eastern European landscapes.

One of the questions I keep coming back to in this book is about how to reveal these ecological scars since, in Eastern European cultural memory, trauma is perceived as more than human. What sources should I choose in order to adequately capture Eastern Europe and its regional environments, where culture, history and memory intertwine with natural resources, and human memory intersects with non-human lives? As I tried to answer this question, my narration mainly became driven by literary sources – works of fiction, and more broadly the text of cultural memory – where imagination participates in shaping the language for reconstructing historical reality. Whenever I can, however, I show how language itself was affected by the Soviet colonization of Eastern European cultures and environments; how it was contaminated by Soviet propaganda and a worldview deformed by communism. The scope of the literature I analyse demonstrates a larger argument: history must be reconsidered through memory of the Stalin era and the late stage of heavy modernity. When we see how the Soviet period can be critical for today's environmental concerns, we recognize that this history is unfinished.

If we define heavy modernity, following Zygmunt Bauman, as the era when 'wealth and power was firmly rooted and deposited deep inside the land – bulky, ponderous and immovable like the beds of iron ore and deposits of coal' (2000: 114), it includes other post-1945 empires, not only that of the Soviets but also communist China or the United States. The late stage of heavy modernity was characterized by grand construction projects such as huge factories, highly industrialized cities, gigantic river dams and canals – and also by objects that got out of control. One could say that the period of heaviest modernity involved the construction of the atomic bomb and the accompanying development of nuclear power plants; it unleashed hyperobjects (on this concept, see Morton 2013) and triggered what is called the Great Acceleration (McNeill and Engelke 2014). Since 1945, the escalation of environmental problems and threats has been proceeding so fast that a shared perspective between East and West – based on comprehensive studies of the extent to which the Cold War, as well as Soviet bloc countries with their unique sociocultural background, were involved in the process of environmental degradation – has not been yet developed. Most historical studies concentrate on political and economic causes and their impact on related environments, while I am interested in contextualizing the environmental cultures of Eastern European countries not only via historical study of the Soviet era, but also through cultural memory as it is articulated mostly in literary texts. These sources structure this book's narrative as an environmentally resistant response. This work focuses on those historical events and cultural phenomena that have shaped and reshaped memory of human relations with the environment, and where nature reveals the gaps in cultural memory or even communicates through them.

The timeline of the book stretches from the period of Stalin's reign (1927–53), which was bisected by the Second World War, through the Iron Curtain period (1945–89) to the dissolution of the Soviet Union (1991). It accentuates distortions in human–nature relations and how they resonate in cultural memory, beginning with events from the period when Soviet communism was established in certain parts of Europe – Stalinist collectivization in the 1930s, various mass crimes buried in the natural landscape, and the development of the extraction and nuclear industries. The book, however, signposts these historical events with ecologically powerful figures from Eastern Europe's cultural memory, including Katyń, Chernobyl and Białowieża, and attends to phenomena specific to the region, such as transformations in its rural and mining cultures.

In Part One, *Unknownland: Retelling the Environmental History of Soviet Eastern Europe through Literature and Cultural Memory*, readers will find a theoretical and methodological toolkit that explains how this book draws on the different languages of literature, history and memory to reconstruct environmental cultures of Soviet Eastern Europe across and beyond individual states' borders. This combination of languages is used throughout the book as a method for reworking historical narration and redirecting the field of memory studies from representing humans as the only victims of war and communism to including nature as a material and imaginary witness and co-mourner for the violent Soviet past.

The environmental cultures of Soviet Eastern Europe involve events and phenomena that are widely represented in cultural memory of the region. I describe them by splicing together my ecocritical reading of literature and historical case studies. They are presented after the toolkit in Part One as follows:

Part Two, *The Tired Village*, addresses the socio-ecological transformation of peasant villages by sustained Soviet collectivization, initiated by Stalin in the 1930s and partly extended to other Eastern bloc countries after 1945. As reflected in rural culture and, primarily, in literature, peasants' responses to collectivization resonate through interconnections among humans, animals and soil exhausted by Soviet intervention. The tired village is the voice of a tired land – in fact, it is a polyphony of voices, rejected at the time and here reconsidered in the framework of cultural memory.

I analyse how Andrey Platonov, in his novel *The Foundation Pit*, refracts more-than-human traumas through language contaminated with propaganda – combining the real destruction of the village with unreal fantasies of successful collectivization. The next chapter begins with propagandistic writings and newsreels celebrating collectivization. I then explore representations of farm animals and trace how pastoral images of the village change over time through readings of Nikolai Zabolotsky's ambiguous *Agriculture Triumphant*, his and Julia Hartwig's poetic portraits of a bull and a cow, and Edward Redliński's novel *Konopielka* about a village's resistance to draining its wetlands. The last chapter of this part analyses both László Krasznahorkai's novel *Satantango* and its film adaptation by Béla Tarr. I show how important the environment and non-humans are to *Satantango*'s representation of a run-down collectivized village near the end of communism.

Part Three, *The Earth's Memory*, deals with sites affected by extensive coal and uranium extraction during communism. Through the example of Szczepan Twardoch's novel *Drach*, I show how the inhuman voice of the Silesian land redirects cultural memory from the region's human history of ethnic and national conflict to its environmental history of coal extraction. I also indicate how the novel depicts the material bond between this region's mining culture and its environmental memory. In the next chapter, through a book of reportage about a town on a mountain in Lower Silesia, I analyse patchy cultural memories of the period of Soviet uranium mining, which are evoked by actual gaps and holes that mining left in the mountain and that tend to open up unexpectedly under buildings and people's feet.

Part Four, *The Persistence of Chernobyl in Cultural Memory*, examines how the catastrophe functions in Eastern European cultural memory and how nuclear risk traumatized memory and prompted a literary and cultural environmental response. The powerlessness of language in the face of the incident is a theme of Christa Wolf's *Accident*. Because of feeling ultimately vulnerable to the consequences of fallout, Soviet propaganda and the prolonged silence of communist authorities, the author reflects on how to regain language for the experience of Chernobyl. In the following chapter, I analyse how Svetlana Alexievich, in her *Chernobyl Prayer*, gives a polyphonic voice to tragedies for humans and nature, and includes the suffering and neglected memory of animals, who are an inevitable part of testimony about the Chernobyl catastrophe.

In Part Five, *Disturbed Landscapes*, I consider natural sites that preserve biological remains and historical traces of the worst twentieth-century genocides and mass murders in Eastern Europe, as they appear in the discourse of cultural memory. However, these landscapes, which are taken to represent the pastoral imaginary of Eastern Europe, have been critically distorted, not only by Soviet and Nazi colonization, but also by Eastern Europe's nationalist politics and anthropocentric practices of memorialization. What interests me is the tension between cultural and environmental memory in the most fraught and compelling landscapes, such as Babiy Yar, Katyń, Volhynia and Białowieża. Mourning with nature de-anthropocentrizes cultural memory, and, as I show through literary examples, in the absence of monuments erected by humans, witnessing in these sites is dispersed among living monuments and people who co-commemorate the victims of the Soviet era.

This line of exemplification aims to unsettle the static narration of Soviet Eastern Europe's environmental history as nothing more than a sustained bludgeoning of natural systems by communist states. The deposits of cultural memory can be continuously reused as a remedy for lack of critical reflection on a shortage of natural resources. However, the basic task of this book consists of listening intently to what has been already stored in Eastern European texts and landscapes of memory through other-than-human voices and testimony.

Part One

Unknownland: Retelling the Environmental History of Soviet Eastern Europe through Literature and Cultural Memory

1

Narrating History across Borders

In this book I examine texts, events and phenomena that help explain the environmental cultures of the former Soviet dominion and reconnect memory and environmental history through literature. Eastern Europe here delimits the spatial and historical context in which all the cultural phenomena described take place. Considering the geopolitical timeline, the book mainly focuses on the period of Stalin's reign and the Soviet domination of Central and Eastern Europe, including the European part of Russia, after the Second World War. In fact, the Soviet era began much earlier, when the October Revolution, also called Red October or the Bolshevik Revolution, established the Russian Soviet Republic as the world's first constitutionally socialist state on 7 November 1917, with the imposition of communist ideology and the execution of Tsar Nicholas II and his family. However, when it comes to nature as an essential part of the Soviet communist project via the state's industrialization, collectivization and militarization, the real starting point is exactly in line with Stalin becoming the dictatorial ruler of the Soviet imperium towards the end of 1927, after the expulsion of Trotsky. To put it even more precisely, the most harmful and large-scale anti-environmental programme was Stalin's so-called Plan for the Transformation of Nature, which was initiated in the 1930s and partially implemented in other subjugated countries after the Second World War (Josephson 2016: 8–36).

The year 1945, following the Yalta Conference, was a breaking point for many distinct countries and regions that were suddenly brought together to form an isolated island – the Soviet bloc – an *Unknownland* for non-Easterners who were luckily excluded from the Soviet phantasm of community. The rise and fall of this vast body of socialist land – 'the land-utopia', as Svetlana Alexievich called it – strained not only countries within it but also the entire non-human world (The Nobel Prize 2015). The Stalinist system, in contrast with that of the Nazis, precluded any environmental protection, although Chernobyl was the first event to open people's eyes to the negative consequences of intensive Soviet modernization.

However, scholars dealing with Soviet history and its environmental legacy tend to concentrate on Russia (Ziegler 1987; Pryde 1991; Feshbach and Friendly 1992; Feshbach 1995; Josephson et al. 2013) and on the extension of Western European political and economic systems following the collapse of the Soviet Union (Manser 1994; Carter and Turnock 1996; Webster 2016). While they criticize the polluting economies of Eastern European countries, their analyses typically ignore the cultural dimensions of this phenomenon.

In order to avoid the simple but still-dominant schema of contrasting East with West, and the stigmatization of the 'Eastern' in Western environmental culture, I reread texts on Soviet Eastern European environmental history in conjunction with the cultural memory and other-than-human voices I find in literary narratives. I ask questions about the criteria that should be used to identify sources for the cultures of nature in this part of the world and seek to reframe the sweeping notion of 'Easternness' that was 'naturalized for the broader Western audience' after the fall of the Iron Curtain (Trojanowska, Niżynska and Czapliński 2018: xvi). Roskin argues, for example, that 'East Europeans do not think of themselves as East Europeans' and suggests that phrases such as the 'Eastern bloc' are controversial (Roskin 2002: 2). While Soviet Eastern Europe can be equated with Warsaw Pact signatory states (Maslowski 2011: 13), the countries thus encompassed developed distinct identities, responses to their political situation, and cultural representations of the environment. Even so, it is possible to discern broad patterns amid the variation and identify some tropes and figures of cultural memory that are linked across Eastern Europe.

To enhance memory of the communist period and extend it to environmental representation, I do not treat Eastern Europe and Central Europe as separate entities, no matter how sharply their cultural identities differ (Ash 1989; Maslowski 2011). Hungary, the Czech Republic and the former East Germany geographically belong to Central Europe, but when it comes to reconstructing environmental cultures in the period of Soviet domination, the histories of Central and Eastern Europe merge. Writers and intellectuals like Milan Kundera, Andrzej Stasiuk and Juriji Andruchovych in *My Europe* (2000) have concentrated on cultural phenomena shared by the whole of (post-communist) Eastern Europe, while historians such as Robert Traba have focused on 'places of memory' (2015–17) spread across this vast region. Environmental cultures should also be considered such hybrid phenomena.

Political borders are obstacles for revising historical narratives to include environmental cultures because they do not admit a variety of non-human witnesses to past events, such as mountains, forests and trees, meadows and ravines, rivers and lakes, or animals. However, these beings, as well the soil, weather, particular landscapes, polluted environments and even hazardous radiation emitted by nuclear waste are the real sources of this *istóriya*[1] – not Eastern Europe's shifting borders. These non-human historical actors prompt me to reflect on the gaps in historical narratives and on the way I reread texts of cultural memory and ask who or what should be included in reconstructing the Soviet past's troubled human relationship with nature.

To some extent historical discourse reflects the structural and conventional guidelines of its leading branches (e.g. political, social, economic, military and even environmental history), because the borders between humans and the environment are themselves historical and we constantly participate in the process of renegotiating them. Many authoritative historians certainly understand history as mankind's activities in the past, as conscious action (Collingwood 1946), or as 'the science of people in time' (Marc Bloch, see Domańska 2006). The same goes for major German philosophers of history who produced the influential historical narratives, such as the history of human freedom (Georg Friedrich Hegel) or the history of moral constraints (Friedrich Nietzsche, Max Weber). Such classics raise neither the question of non-human histories nor the natural status of humans in the course of history (Chakrabarty 2009: 214). In other words, the historiography of non-human actors and the problem of environmental representation is either excluded from human-centred historical narration, or subordinated to explaining human involvement in past events.

However, some useful ideas about how to challenge anthropocentric historiography have emerged via a sceptical approach to historical research. The French historian Paul Veyne argues (1984 [1971]) that history has neither a clearly defined object nor a particular method. Historians study not only people but everything that is specific and took place in the past (Veyne 1984: 59). The only restriction historians impose upon themselves is that they do not search for laws, which they leave to social scientists. There is not even a consensus about a minimal definition of history as a discipline concerned with 'humankind in the past':

> Then what is history? And what do historians, from Thucydides to Max Weber or Mark Bloch, really do, once they have gone through their documents and proceeded to the 'synthesis'? Is their work the scientifically conducted study of the various activities and the various creations of men in other days? the science of men in society? human societies? . . . The science of what sort of societies? The whole nation, even humanity? A village? At least an entire province? A group of

bridge players? ... The human presence is not necessary for events to rouse our curiosity.

Veyne 1984: x, 3, 58

I quote from a passionate essay in which Veyne concludes that humanity is too narrow to serve as an historical object, since historical events without 'human presence' still intrigue us and, I would add, some histories cannot even be reconstructed without considering other-than-human actors. In a broader perspective, human history must be combined with that of geology and environments because they are not only related but challenge each other. As Dipesh Chakrabarty argues in his critique of anthropocentric historiography, *The Climate of History*, we face the collapse of the 'age-old humanist distinction between natural history and human history' (2009: 201). The global climate does have a history (Carey 2017), as do forests (Grewe 2010; Brain 2011; Brock 2017). Including non-human animals in historical narratives is actually necessary – for example, to describe the lives lost in the First World War (Baratay 2013) or to write the history of the circus, in which animals have obviously long been central performers (Neirick 2012).

Writing histories that include non-humans, or even writing histories of non-humans, may shed some light on historical scholarship's approaches to writing itself. In this case, an historian is like a writer who enters unknown territory with no definite borders and searches the most convincing language for characters who are under-represented in historical discourse and that, like climate or soil, need not be anthropomorphized to be included. Ecocritically-oriented historians, informed by theory, focus on how to de-hierarchize historical studies by recognizing the material and cultural status of non-human beings, entities or phenomena, by reconstructing how they are remembered and forgotten, and by developing adequate strategies for de-anthropocentrizing historical narratives. If, since ancient times, history has been perceived as a powerful didactic tool – people can learn everything from history (Cicero) – then the way we write and tell histories really matters in a time of environmental crisis.

Different kinds of boundaries – historical, political, geographical and disciplinary – shape this book's historical narrative about Eastern European environments and their cultures. As such, my effort to narrate this history involves drawing on philosophy to adapt language for crossing anthropocentric borders and also rereading sources that can help us understand Soviet Eastern Europe's *unknownland*. Finally, it involves experimentation with methods for combining historical writing with analyses of literary and cultural texts, in order to connect the region's environmental history and its cultural memory. The variety of cultural sources I use to contextualize the ecological complexity of the Soviet past call for more explanation about how literature, history and memory intersect in this book. This is further discussed in the following three chapters: History and Literature, Environmental History, and Cultural and Environmental Memory.

Where are the gaps in existing historical narratives of this period and how can they be filled to address environmental cultures in Soviet East Europe? To question Eastern Europe's stigmatization as anti-environmental, we need to examine the ecological scars left by communist governments in this part of the world by paying attention to those most affected archives.

2

History and Literature

History has long been influenced by literature. Indeed, literature's influence on history has been quite well discussed in historiography in terms of both the function of literary sources – especially biographies, letters and diaries – as documents and the literary style that historians employ to create their own narrative. Many of these issues arise in dialogue between historians and literary scholars when they are asked a general question: what does literature offer history? This book answers by showing how literary sources that represent environments in Eastern European cultures can inform historical discourse.

In this chapter, by referring to a discussion conducted in Poland by two generations of scholars, I reconstruct arguments for a multi-functional, epistemological and methodological approach to literature when coupled with historical work. The matter under discussion is whether literary work can 'appropriately saturate a historian's knowledge with non-source knowledge' and become 'an inspiring source for interpretative techniques'. The author of these words, the historian Jerzy Topolski, ascribes significant cognitive content to literature when he writes about historians' use of literary texts and argues that literary writers have the ability to 'concentrate attention' around the key features of reality, primarily through the use of creative narrative techniques (1978: 12–17). In turn, Krzysztof Pomian, also an historian, suggests that 'historians unwittingly create fictions by using the imagination' because historical narrative can never take the form of a pure report (2014: 34). Przemysław Czapliński, a literary critic and theoretician, goes even further: although he makes Polish literature his starting point, he formulates a thesis about 'the fictional methods of reaching knowledge about the past' (2014: 283). His understanding is in line with the position of another literary scholar, Kazimierz Bartoszyński, who states that a literary text, taken as a source, must be recognized as an element of a larger whole that can be called a pragmatically understood situation of communication. We take into account not only the message conveyed by the text, but also the medium (channel) through which it is transmitted and which is 'the information itself' (Bartoszyński 1978: 61). This mediatory role of literary texts is possible because of literature's 'informative abundance (in a semantic sense)' (61). Following the American narrative turn, which reached Poland much later, Bartoszyński writes about an open and flexible formula for historical narratives – one that is, like literature, based on 'non-finality' and on 'affiliation to the culture of disclosure' (63). It means abandoning closed narratives that simply try to describe historical events and actions in favour of formulating 'attempts' to recreate psychological perceptions of experienced events (Pomian 2014: 22). This discussion ends with a question posed by Michał Głowiński, a historian of literature, about whether historical knowledge can be gained by reading literature (2014: 95). According to him, we can learn about history by reading literature when we pay attention to specific properties of a literary work, which make the text literary, such as stylistics and poetics, usage of metaphors and artistic language; then it can be read historically as a message containing ideas, a testimony of certain possibilities and ideological potentialities (Głowiński 2014: 104), which are part of cultural memory. In an earlier essay, *The Novel and Truth* (1973), he says that only literary utterances can make historical events more real.

What makes this debate relevant is, on the one hand, its intergenerational and interdisciplinary coherence in acknowledging the intertwined roles of literature and history and, on the other, its methodological potential for hybridizing the language used to describe past environmental cultures by interlacing fiction with non-fiction. Moreover, literary style can not only filter into historical narration and help in describing the past, but literature can also serve as an epistemological source for historical knowledge and prompt possible reconfigurations in memory, especially in the absence of human witnesses, as the example of Katyń Forest will show.

Historical writing and the construction of historical knowledge can be, however, even more closely bound together, and the division between literature and history can be presented as even less clear. This is the case with historians who radically transgress the border between historical and literary language and become renowned specifically for emphasizing the role of literary narrative in world historiography – namely, Frank Ankersmit and Hayden White.

The relation of history and literature is the main problem Hayden White analyses, alongside the question of boundaries of what he calls historical prose as narrative discourse (1973: ix). The radicalism of White's proposal consists in equating historical and literary writing, since to think historically means, not to explain, but to imagine historically (White 1973: 2). The common ground between historical and literary narratives takes the form of 'an extended metaphor': 'the historical narrative does not reproduce the events it describes', although it 'tells us in what direction to think' or 'calls to mind images of the things it indicates, in the same way that a metaphor does' (White 1986: 91), and there are a variety of literary strategies for doing this. Finally, White concludes that:

> history – the real world as it evolves in time – is made sense of in the same way that the poet or novelist tries to make sense of it, i.e. by endowing what originally appears to be problematic and mysterious with the aspect of a recognizable, because it is a familiar, form. It does not matter whether the world is conceived to be real or only imagined; the manner of making sense of it is the same.
>
> 1986: 98

In another text, he holds that fiction does not exclude epistemological functions: every written form is cognitive in its aims and mimetic in its means because it represents someone's experience. Referring to Primo Levi's *If This is a Man*, Art Spiegelman's *Maus* and Roberto Benigni's *Life is Beautiful,* White argues that there is no sense in rejecting these artistic testimonies as unhistorical only because they are fictional or use visual aesthetics; after all, they are about an actual historical event – the Holocaust (White 2014: 28–9). As a result, White treats factual communication as a subset of fictional writing. Though White's position is extremely constructivist, his work belongs to a genre that is worth experimenting with. The literary historical narrative can theoretically reframe anthropocentric versions of history and include, via fiction, non-human actors, if such a hybrid text combines historical and literary scholarship.

Frank Ankersmit presents views similar to those put forward by White. In the first of his 'Six Theses on Narrativist Philosophy of History', Ankersmit holds that 'historical narratives are interpretations of the past' (1994: 33). Certain historical schools decide what constitutes historical narratives and how to interpret them. Put another way, the historian only 'represents' the past. Reality 'does not require that the past itself have a meaning ... Meaning is originally representational and arises from our recognition of how other people (historians, painters, novelists) represent the world. It requires us to look at the world through the eyes of others – or, at least, to recognize that this can be done' (102).

Ankersmit's recipe for writing history involves incorporating literature as a source of meaning in a weak sense. Making history by looking 'through the eyes of others' prompts the question of whether this human community of perception can be extended to non-human eyes. We cannot radically change our (human) perspective, but the question of *how cultures make environments speak* includes literature as a source of historical knowledge. As a result, the historical logic of cause and effect of an event such as the rise and fall of the Soviet Empire ceases to play a central role because what literature does in this book, is substituting the anthropocentric framework of explaining the past with reading environments and asking what do they communicate when they are placed centre stage in a historical narrative. In this book, Soviet politics takes a back seat to environmental damage in a story about broken ties between people and nature and their poetic reconnection in cultural memory.

3

Environmental History

There is a deeply engrained cultural conviction that nature exists outside history, in a timeless realm where memory does not operate. Idealizing a transcendent nature and separating it from an anthropocentric account of the past contrasts with the reality of the contamination and simplification of our biosphere. Environmental history, as a discipline, has radically shifted such assumptions and drawn nature into its historical analysis. It provides a useful orientation to events that characterized the destructiveness of the Soviet regime and the whole communist part of the world – though perhaps these states were, in environmental terms, no more harmful than their capitalist counterparts (Bruno 2016: 20).

I do not aim to polemicize for or against an unquestionably prevalent way of presenting Soviet environmental history as the record of how a centrally managed economy implemented its socialist hypermodernization project to make the Soviet Empire independent of the Western world without calculating the costs for people and nature. Environmental history, however, is dominated by English-speaking scholars, both as founders of the discipline and representatives of Anglo-American academic institutions (Isenberg 2017). These scholars address global, continental or regional phenomena, including the history of Eastern Europe, as in the case of research publications about the Cold War (Blackbourn 2006; McNeill and Unger 2010; McNeill and Engelke 2014). Environmental studies of the Soviet period fit in with global environmental historical discourse, but some scholars do indeed complain that dominant interests in the field relate to the anglophone world and that most publications are in English (Moon 2017). Perhaps this is why studies of Soviet Russia's environmental past tend to contrast East with West and present 'the Soviet Union as an environmental-catastrophe state' (Moon 2017: 36). At the same time, representations of the events on which these accounts of environmental history are based are more or less the same. They form 'a litany of Soviet environmental disasters' and are presented:

as if part of a moral lesson for the rest of world of what can go wrong without due care for the natural world: massive pollution or degradation of the land, air and water, and harm to human health, as a result of rapid, centrally planned industrial and agricultural development; the conversion of free-flowing rivers to chains of lakes by big dams; the threat to the pristine waters and unique ecosystem of Lake Baikal from cellulose plants built on its shores; the drying-up of the Aral Sea in Central Asia as a result of the diversion of waters from the rivers feeding it to irrigate the surrounding arid lands to cultivate cotton; an aborted plan, following warnings from scientists, to divert rivers from Siberia and the Far North to replenish it; and above all the catastrophic explosion at the Chernobyl nuclear power station, as a result of design flaws and a test that went disastrously wrong.

Moon 2017: 35–6

Such critiques notwithstanding, the list of environmental catastrophes in the former Soviet Union that affected both other Soviet bloc countries in Eastern Europe and nations in Western Europe is indeed very long (Brain 2012). It is almost impossible to recount them without severely criticizing Soviet leaders, their utopian, anti-environmental vision of nature as a mere resource and their related politics of aggression against the non-human world (Josephson 2016). However, there have been attempts, including by leading environmental historians of the Soviet Union, such as Douglas Weiner and Paul Josephson (Bruno 2018: 148), to narrate such events outside of this discourse of stigmatization and Western self-congratulation.

Weiner, for example, describes pioneering conservation practices in Soviet Russia which were in opposition to the environmentally harmful politics of party leaders. He devotes a lot of space in his *Models of Nature* to the development of a network of *zapovedniki* (nature reserves that also served as scientific laboratories, where fishing and hunting were restricted or prohibited), natural parks and monuments of nature (places dedicated to protecting rare, exceptional or valuable ecological species and types of ecosystems). The law establishing these three types of protected areas was issued under Lenin's rule in 1921 (Weiner 2000: 27). Weiner examines key conservationist journals, including *Okhrana priridy* ('The protection of nature'), which – until it was censored in 1931 – criticized the first negative ecological consequences of collectivization. This journal published research results covering pollution and damage to the environment even during the harsh period of Stalin's rule, when there was a high risk of being persecuted or even executed for writing against the propagandist media's version of events (Weiner 2000:

121–3, 144–6). Therefore, as Weiner and others have shown, the conservation movement was active in the USSR, but it was considered an enemy of 'socialist construction' (Weiner 2000: 146) and was severely suppressed during the Stalinist regime.

Josephson, who considers the Soviet Union's dominant exploitative approach to nature to be similar to that of capitalist countries (2013: 12–13), emphasizes that the differences between socialist countries and the West lie in political conditions for coping with environmental problems: 'people under socialism were largely silenced by their leaders from speaking openly and actively about environmentalism' (2013: 3). Other historians hunt for positive examples, such as that the Soviet Union planted the most forests in the world between the 1930s and the 1950s, as Stephen Brain (2011) has shown, or that the pollution of air and water in the former Soviet states was no worse than in the United States (Bruno 2018: 149–50). The production of hazardous spaces because of toxic waste and heavy metal effluent is considered to be a concrete legacy of Soviet communism (Bruno 2018: 149), more than environmental damage that led to stigmatizing and 'exceptionalising' generalizations about the Stalin era (Bruno 2018: 150). Above all, recently historians have disputed the theory of the USSR's ecocide that Murray Feshbach and Alfred Friendly developed in 1992 (Bruno 2016: 18–21). They have also proposed a different periodization for assessing the environmental situation that not only accounts for the destructive impact of the Stalin and Khrushchev eras, but also that of the 1970s and 1980s (Bruno 2018: 149–50). The discussion fluctuates, but the lack of scholars from post-Soviet countries participating in it has not significantly changed.

It might seem that a very basic definition of environmental history as a research field can be shared across geographically diverse areas, since we are dealing with a type of historical scholarship that refers to current environmental issues by analysing their historical background. According to one of its journals, environmental history is an interdisciplinary field that addresses 'issues relating to human interactions with the natural world over time, and includes insights from history, geography, anthropology, the natural sciences, and many other disciplines' (*Environmental History* journal website). The editors of *The Companion to Global Environmental History* offer another handy definition: environmental history 'is the study of the relationship between human societies and the rest of nature on which they depended' (McNeill and Mauldin 2012: xvi). The idea that the physical environment has affected civilizations was first adopted by historians of the Annales School to describe *longue durée* developments that shape human history (Bramwell 1989: 40–1).

While the majority of (environmental) historians tend to study the environment as an extrinsic historical factor that provides an increased understanding of human actions, recent environmental historians have gone further and claimed that history is not just 'human'. On the contrary, 'history is inextricably linked to natural agents of change, both organic and inorganic' (Breyfogle 2018: 14); the natural world participates in past events, not passively but actively, through human and non-human subjects (Bruno 2016: 8). To address the natural agents of history, new methodologies are brought into historical research based on fieldwork, interactions between historians and nature, and interdisciplinary projects involving historians and scientists (Breyfogle 2018: 17). Interdisciplinarity, mostly in relation to the environmental sciences rather than cultural studies, has in fact been postulated since the discipline started to be recognized as far back as the 1980s (Isenberg 2017: 4), the 1970s (McNeill and Mauldin 2012: xvi), or the 1960s with the publication of Rachel Carson's *Silent Spring* in 1962.

When it comes to periodization of the discipline, environmental scholars have divergent opinions. Some state that environmental history was first written by nineteenth-century geographers who stressed the influence of the physical environment on the development of human society (Sörlin and Warde 2009: 4–5), but the chronology of the discipline is a complex issue: 'It is neither a recent and merely political feature, as suggested by the "1962" (Carson) chronology, nor a discursive invention of "1864", the year of George Perkins Marsh's sweeping yet highly influential geographical tract *Man and Nature*, where man was elevated to the dubious role of principal agent of physical change' (Sörlin and Warde 2009: 9).

All in all, environmental history is a dynamic and widely debated field driven by contemporary concerns. It includes very technologically advanced, interdisciplinary analysis, such as empirical science and GIS mapping, as well as more general study of human exploitation of the natural world. Yet the field's mission exceeds scientific parameters such as those of historical ecology. As two major environmental history associations, one American and one European, indicate, the field aims not only to foster greater understanding of 'human interactions with the natural world' and facilitate dialogue among humanistic scholarship, environmental science and other disciplines, but also to support global environmental history efforts that benefit the public (ASEH.net; ESEH. org). Reaching the general public and participating in policymaking and education are goals shared with other organizations representing the environmental humanities and related subjects, including ecocriticism (ASLE. org; EASLCE.eu).

Building a connection with ecocritical scholarship via what is recognized as the environmental humanities (Bergthaller et al. 2014: 263) might increase the likelihood of changing the environmental historical narrative about Soviet Eastern Europe. There can be no doubt that environmental historians, as well as other representatives of the environmental humanities, are ahead of contemporary debates about environmental problems and climate change. Their work involves major recapitulations, such as *The Great Acceleration* by J. R. McNeill and Peter Engelke (2014), as well as finely tailored projects like Andy Bruno's arctic environmental history (2016). The period of the Cold War is an irreducible part of this discussion, mostly because the fossil fuel extraction, agricultural intensification, and nuclear and hydropower industries (McNeill and Engelke 2014: 18–21, 25–33) that took off then are central causes of current global environmental problems. However, recognizing non-human actors in historical studies of the Soviet environmental heritage (Bruno 2016, 2018; Breyfogle 2018) opens narrative parallels with the historical reconstruction of environmental cultures, which I draw from the repository of cultural memory.

Bruno states that his monograph on arctic environmental history is the first to 'fully consider alive and inert elements of the natural world as participants in the dramas of Soviet history. Animate and inanimate materials were not just passively acted upon as objects, but also played a role as subjects in this story' (2016: 8). For example, migrating Kola reindeer interfered with Soviet programmes of environmental colonization (2016:10). Bruno refutes anthropocentric constructs of nature as the passive material of history and incorporates into his research theoretical ideas about reassembling humans and non-humans when reconstructing the past of Soviet Eastern Europe. Much like his approach to history, the present narrative also focuses on non-human participation in past events and the ways in which human interactions with nature were perceived by rereading relevant literary sources. Non-humans, including rocks as well as animals, are being rediscovered as subjects of history (the role of rocks in Soviet history is presented in Bruno 2018). Non-humans' historical records are still scarce, and they cannot be treated as a well-established, autonomous voice in history because they are not sufficiently or adequately represented, no matter how actively they participated in the past, not metaphorically but actually. Against this background, therefore, when conventional historical archives lack the sources, perhaps there is an argument for mobilizing different approaches to representing non-humans not only as actors in history but also as shapers of historical and literary narratives. I am particularly interested in how Soviet Eastern European cultures activate non-humans in memory to empower narratives of environmental resistance to communism.

Asking questions about how cultures accommodate or suppress non-humans in stories about the past prompts me to search for new approaches to writing the history of environmental cultures and to include literary texts and other cultural resources as part of historical narrative. In environmental history, literature can motivate such reflection. Thanks to its fictional character, literature can scale or reorder the presentation of facts; ask questions about the language that should be used in describing historical relations between humans and non-humans; illuminate historical events in which the bonds between humans and nature were damaged; and, from an ecocritical perspective, strengthen the representation of non-humans as historical actors by limiting the human exceptionalism. This book's narrative reconstruction of Soviet Eastern European environmental cultures is driven by an effort to speak on the behalf of communities beyond the human and recognize them in the work of cultural memory.

4

Cultural and Environmental Memory

Historical discourse, including the environmental – or 'anti-environmental' – history of the region, does not exhaust Eastern European cultures of nature when considered under the auspices of memory. Admittedly, core events of the Soviet era have already been extensively analysed in environmental history, in terms of politics and economy, and in their various shades and contrasts in each of the former Soviet bloc states – for example, different models of collectivization in the Soviet Union and its Eastern European satellites have been discussed (e.g. Bell 1984). Such historical events, though, only provide a static background for the dynamic process of forgetting or remembering a traumatic past. These events resemble twines or knots that memory cinches tighter, unties or even cuts.

The process of collectivization in the Soviet Union during the first Five-Year Plan was catastrophic; it resulted not only in massive famines that killed millions of people, but millions of animals also starved or were slaughtered in revenge, in acts of defiance against unfair Soviet reforms. Chernobyl is remembered because it has become a cultural symbol and a referential sign for the collapse of the Soviet era. But the Chelyabinsk nuclear accident, also called the Kyshtym or Mayak disaster, is largely forgotten, despite convincing historical studies which show that it was similarly catastrophic in polluting vast areas of the environment and human settlements. Chernobyl, too, should be remembered not just through stories of the people who sacrificed themselves to contain the meltdown in feats of legendary heroism, but also through stories of killed and abandoned animals, destroyed forests and trees felled in their thousands. Although these non-human losses are recorded in terms of numbers, they are marginalized in historical reconstructions. There are, however, powerful literary accounts in which human and non-human histories complement each other. This book seeks to fill gaps in cultural memory by retracing the region's environmental cultures, examining those examples that activate environmental memory by bridging history and literature.

Contemporary debate over memory, which is identified as the field of memory or cultural memory studies, has the potential to contest or even intervene in historical discourse that is dominated by an anthropocentric perspective (Kennedy 2017: 268). Trauma studies and affect theory have significantly catalysed the multi- and interdisciplinary field of memory studies in relation to East European history, especially in reference to the Holocaust (Nikulin 2015: 22), and intensified the search for means of expression beyond historical discourse to represent collective experiences of trauma. Pierre Nora stressed the 'spontaneous' and vitalistic character of memory as 'the phenomenon of the present' in contrast with the 'rational' and ordered past (Nikulin 2015: 12), and assigned carriers of this real memory to a spatial but metaphoric category of 'memory sites' (Nora 1984–92). Other theorists of memory studies, such as Jan Assmann (2006, 2011), Aleida Assmann (2012), Michael Rothberg (2000, 2009) and Astrid Errl (2011 [2005]), have continued to identify a variety of memory media, including material objects, places, and immaterial works of fiction and art. In particular, Jan Assmann's definition of 'cultural memory' appears to be ontologically 'text-based', since its meaning is circulated in 'writing' (Assmann 2011: 8): that which can be remembered is stored in the form of a text (Nikulin 2015: 15).

Memory studies and the study of environmental cultures also potentially intersect because theoreticians of memory are not historians as such, but theorists of culture. The field of cultural memory originated both in the act of separating memory from history (Assmann 2011: 29–30) and in efforts to show how memory is able to construct 'new histories' (Nikulin 2015: 5). 'Cultural memory', which is sometimes used interchangeably with 'collective' and 'social memory', is certainly a multifarious notion. The term is often used in an ambiguous way, without providing any rigid definitions, especially by the founders of this intellectual field, i.e. Maurice Halbwachs, Pierre Nora and Aby Warburg. Media, sites, practices, texts and structures as diverse as myths, monuments, places, historiographies, rituals, images, symbols, configurations of cultural knowledge and neural networks are now subsumed under the wide umbrella of 'cultural memory'. This explains why so many scholars reconstruct Halbwachs's or Nora's theory before they use the terms 'cultural memory' or 'memory site'.

Because of its intricate character, cultural memory has been a highly controversial issue ever since its conceptualization in Halbwachs's studies of *mémoire collective* (1925, 1941, 1950). Halbwachs was, however, the first not to attribute memory to humans' internal abilities to remember information stored and transformed in the brain, mind or consciousness. In contrast to empirical approaches to such processes, he located memory in 'external' sources that are part of 'historical cultural studies'. Therefore 'the contents of this memory, the ways in which they are organized, and the length of time they endure are for the most part *not* a matter of internal storage or control, but of the external conditions imposed by society and cultural contexts' (Assmann 2011: 5). What is more, Halbwachs's understanding of 'collective memory' ties it to 'the present purpose of a group'; therefore, 'collective remembering' is not fixed but rearranged to establish 'the meaning of the shared past'. While memory circulates within a group, it also fluctuates (Assmann 2011: 29) and is not standardized as in the case of historical, universalized memory (Nikulin 2015: 11). In reference to Halbwachs's *On Collective Memory*, Assmann recapitulates that memory 'works through reconstruction. The past itself cannot be preserved by it, and thus it is continually subject to processes of reorganization according to the changes taking place in the frame of reference of each successive present' (Assmann 2011: 27). This distinction between what belongs to memory and what belongs to history is quite radical, yet it gives us some idea of Halbwachs's influence in differentiating memory as a flexible, living cultural phenomenon that is linked to culturally variable contexts – a concept that would later be further developed by others, starting with Pierre Nora.

However, the expansion of memory studies from a cultural focus to interdisciplinary contexts was poorly grounded in a separatist rejection of history and its limits of expression, as Halbwachs and Nora characterized it. In reality, their understanding of memory studies resonated with internal transformations of historical discourse and even the linguistic and cultural turns discussed by historians like White and Ankersmit. Therefore, the boom in memory studies was not the result of a crisis of history (Lee Klein 2000: 143), but of geopolitical changes such as 'the end of the Cold War', when 'the binary structure of eastern and western memory cultures ... collapsed' (Errl 2011: 4). The spread of popular media and technologies and poststructuralist transformations in historiography also played a role (4–5). As a result, according to Errl, memory has turned out to be 'a transnational phenomenon' (2011: 4).

Memory studies have been criticized from the beginning for transferring concepts from individual psychology to the level of the collective and for not relying on well-established studies of tradition or cultural heritage. Conservative critics have argued that there is no need for additional research, and that history does not need memory (Collingwood 1994 in Nikulin 2015: 12 n. 21). Conversely, if 'cultural memory studies' involve 'the interplay of present and past in sociocultural contexts' (Errl 2008: 2), then cultural memory, as a discipline, moves beyond conventionally defined subject areas and disciplinary boundaries and locates itself at the intersection of theory and criticism as well as that of literary and political practice. Ecocritics (Buell 2017; Kennedy 2017) have not overlooked this relatively new usage of 'memory', which deals with the cultural context of a dynamic process of remembering. This process not only selects and gathers past phenomena for the present situation and adapts them to new problems, but also struggles with a variety of 'cultural' obstacles: 'manipulation, censorship, destruction, circumscription, and substitution' (Assmann 2011: 9).

The traumatizing forces of Eastern European history have deeply wounded human memory, which has become aggressive towards nature in turn. Memory is processed through remembrance and commemoration practices, or what in Poland is called 'memory' or 'historical politics', including the role conservatives ascribe to institutionalized memory as a buttress for national identities, along with a perception of the environment as a subordinated reflector of 'national self-images and stereotypes' (Buell 2017: 110). The dynamic and active field of cultural memory offers the potential for questioning those 'obstacles' pointed by Assmann in historical narratives; this has resulted in recent efforts to build upon memory studies' 'success' in the context of ecocritical theory and to initiate 'eco-memory' studies (Kennedy 2017: 268). Their aim would be to reframe the history of the environment and rehabilitate natural landscapes' or animals' presence, in order not only to narrate human tragedies, but also to express ecological concerns. Such efforts might be called 'environmental memory', understood as 'the possibility that literature and other expressive media might act as carriers of environmental memory' (Buell 2017: 96). In the light of this, reconstructing environmental cultures can be seen as 'a memory-(re)construction project' which operates within the 'environmental imagination' (Buell 2017: 99) by rereading cultural and textual media.

Buell's ecocritical approach states that now is the moment to locate potential sources that may correct memory (Buell 2017: 113), including literary and imaginative sources. If cultural memory can be perceived as 'ongoing work of reconstructive imagination' (Assmann 1997: 14, in Nikulin 2015: 25), then more effort should be made to represent not only memory about humans but also about human–environment relationships and environments as figures in memory. Even as it criticizes the past, memory should also play an active part in reflecting on global survival in a time of severe climate and environmental crisis.

For ecocritical studies, this research on memory, and how memory can revise historical narratives and remind us of what was lost or forgotten, is exciting and timely. It prompts reflection on how we, as ecocritics, might use literary texts to extend cultural memory more consistently to environments and non-humans. Active cultural memory, in its narration, requires accompanying sources as 'alternative epistemologies' to develop (Lindbladh 2008a: 5). Scholars who study memory in literature stress that this imaginary aspect of 'knowledge' is emotional, non-linear and pre-cognitive. Its messages must be extracted and deciphered, as in Andrei Voznesensky's poem *After the Tone* (1991), from the collection *Aksioma samoiska* (1990), about an answering machine that acts as a 'memory carrier':

'I am an answering machine.
You have one minute at your disposal.
Please leave your message after the tone.'
…
'Record this: I am the voice from beyond,
an information leak from the other side.'
…
'Is that Voznesensky's answering machine?'
'This is Tariverdiev the composer's answering machine.
It's verdict time for our owners.
Long live international solidarity for answering machines.'
…
I am an answering machine.
I answer
from the Black River to Kamchatka
for the mutilated age of eternal questions,
for radiation in packets of tea…
for Parnassian porn, for background noise,
Talmuds, Buddhas, Christs, Judases.
You have one minute at your disposal.
Leave your message after the tone.
After the tone.
Blind
chase AFTER
blindchaseafterblindchaseafterblind
chase intone after THE TONE

In this poem, an organized but inhuman and automatic system for recording messages uncovers their absurdly variegated character when they are listened to one after another. The answering machine is a container of every kind of information: work matters mingle with private messages as well as with unrecognized strangers' words and absurd jokes. The messages form a chaotic stream of semantic scraps from different contexts.

The poetic imagination, however, animates this unordered space of dangling, drifting notifications when it makes the answering machine conscious and has it interfere with the chaos. In the poem, the answering machine begins to operate like a self-learning system – proto-AI – which gains its own agency when it says: 'Record this: I am the voice from beyond, / an information leak from the other side.' In a revolutionary – and, of course, socialist – way, the answering machine spreads this message to others, taking charge of the machines to transmit a political message: 'This is Tariverdiev the composer's answering machine. / It's verdict time for our owners. / Long live international solidarity for answering machines.' Normally, voicemail simply saves all messages, without any filter – it is an automaton, a form of storage. Here, it is a device of poetic intervention, when it becomes a form of voiced memory as the poet's machine eventually 'answers': 'from the Black River to Kamchatka / for the mutilated age of eternal questions, / for radiation in packets of tea' (Voznesensky 1991: 65). And it is a surprising device reporting what has been stored within memory of culture poetically intersected with memory of the Far East Russian environments.

'The poetics of memory', a term coined by Thomas Wägenbaur in 1998, 'defines memory on the one hand as storage and on the other hand as a story' (Lindbladh 2008a: 6). In this poem, the chaotic, polyphonic realm of storing messages can be revolutionized by the power of voices to emerge and reconnect with those of other answering machines (Mikael Tariverdiev was a contemporary of Voznesensky and composed music to some of Voznesensky's poems). The answering machine's poetic but non-human, mechanistic voice sounds rebellious in its message from the other-than-human side, and it is not anonymous: it speaks from a poetic Soviet space.

The poetic voice is like an answering machine that receives tones and different kinds of information, but only transforms part of them into a memory as it captures, filters, reorders and forms them for expression. Otherwise, even the most important message evaporates. Literature has the power to pick up an item of communication and make it meaningful and memorable. It can be compared to an ecological force resiliently operating within the broader system of cultural memory and environmental history. This is how Hubert Zapf approaches literature in terms of cultural ecology (*Kulturökologie*) (2001: 85).

In 'culturecology' – defined as adding value to literary texts in the shape of a symbolic function that corrects things deformed and transformed in other non-fiction languages, thereby affecting culture – fiction plays both a meta- and a counter-discursive role. In this sense, it functions according to the old rules of critical and utopian theories. And yet, beyond these two functions, Zapf points to a third function that is much more innovative when it comes to understanding environmental culture. In this third variant, literature is a 'reintegrative inter-discourse' that becomes a component of the whole system of cultural discourses and helps redefine cultural centres (2001: 93). As a result, through fiction it is possible to combine different non-fiction languages, such as historical narratives and memory studies, that often lack ecocritical context. Literary texts can thus interact dynamically and work against stagnant and reductives representations of environments and non-humans in a cultural system, understood here as a whole series of memory carriers, media and processes. In this respect, literature is processual, complex and interdependent, much like any ecological system. Thus, literature connects with what we know from other discourses such as history and contributes to an environmentally-oriented version of Soviet Eastern Europe cultural memory by representing non-human aspects of the past in its literary mode of expression. In this book, literature, as a reintegrative inter-discourse among other discourses, reorders historical narratives and functions as an equal but resistant voice in retelling the Soviet Eastern Europe past through relations with nature. This book is thus a part of a broader project to find space in our human-centred memories for environments and non-humans, to revisit the most painful traumas of the Soviet period and to discover where cultural and environmental memory meet and where they separate.

Part Two

The Tired Village

1

Historical Background

In much of pre-Soviet Eastern Europe, the peasantry comprised the majority of the population, while the bourgeoisie – as a social class that elicits a type of cultural world – had not developed significantly in a socially stratified landscape. Historically, peasants lived under two differing extreme orders: enslaved serfdom lasted until the nineteenth century (and was especially severe in Russia), and later industrial farming enterprises and their barons took over rural areas. Until Stalin began collectivizing the villages in the Soviet Union in 1928, the romanticized myth of the peasant living close to nature, in an organic and cooperative bond with it, was still culturally alive. Socialists brought 'civilization', which started with the rapid modernization of villages and met with both opposition and blind acceptance of the new order. Generally, peasants were uneducated and did not necessarily understand the changes taking place, but they were strongly attached to their traditions and lifestyle. The post-war effort to rebuild and expand cities and towns, and the influence of Cold War competitiveness on the socialist economy, rapidly changed Eastern European villages.

If intensive agriculture, meaning large-scale grain monocultures and unsustainable meat production, has been a decisive factor in the environmental degradation caused by human civilization around the world, then the Soviet period can sharpen 'our understanding of agroecological history' (Breyfogle 2018: 12). For some, agriculture is a key inflexion point in global environmental history because it involves an anthropocentric order of controlling and destroying natural resources without trying alternative, more sustainable models to ensure our survival as a species (Morton 2018). This model of agriculture began in ancient Mesopotamia and 'eventually required industrial processes to maintain [itself], hence fossil fuels, hence global warming, hence mass extinction' (Morton 2018: 10). Alongside transport, industry, nuclear power, waste management, species extinction and global warming, agriculture is thus one of the main issues that raises environmental concerns today, and is also an important driver of global environmental change which needs to be mitigated. Better agricultural practices could solve some environmental problems.

While sustainable agriculture is usually discussed as a goal for the whole of Europe, the production of meat over the last fifty years has almost doubled (Ritchie and Roser 2019). However, the reasons for this increase in meat production may differ in Western versus Eastern European countries due to their cultural and political contexts. For example, in Eastern Europe, meat was a luxury reserved for elites during the communist era, whereas now it is cheap, popular, consumed on a massive scale, and an important part of celebrations such as weddings or funerals. The Soviet period made cultural patterns of diet less environmentally friendly by accelerating meat production (Belyaev and Pilnyak 2019 [1936]; LeBlanc 2016), although we know from historical food studies that, before communism, peasants prepared meat dishes only for major festivities or when they had an animal ready for slaughter. The dominant components of their diet were grains, dairy, and tuber vegetables such as potatoes. During communism, those who had family in the countryside could benefit from having a sausage or a cutlet on special occasions.

Before communism, Eastern European peasants represented sustainable agriculture, which explains why they were considered especially attached to nature. This was a widespread theme in rural culture because they were the only group in society that depended on nature in their everyday lives and so were materially connected with rivers, lakes, soil, forests and animals. Peasants were also forced to adapt to the changes that the Soviet Union imposed – to a greater extent in Russia and Ukraine and a lesser extent in Poland, Hungary and Czechoslovakia. The peasantry in Russia represented the overwhelming majority of the population; peasants were the main victims of Stalin's actions (Heller and Nekrich 1986: 232). Their experience of state intervention and collectivization – so deeply internalized in Ukrainian national identity, as Oksana Zabuzhko recollected (2017) – resulted in major changes to villages, which had been culturally independent from politics. For example, Ukrainian villages were self-sufficient before collectivization because they had the most fertile land in Europe, i.e. *chernozem* (Zabuzhko 2017: 46–7). That was about to radically change, along with peasants' relationship to the environment and indeed the whole discourse of peasantry as a culture linking nature and society, expressed via folk traditions and the highly religious character of the community. Collectivized agriculture brought not only new machinery but also fertilizers, erosion and soil pollution (Josephson et al. 2013: 73). In Ukraine, environmental damage was accompanied by the losses of humans and animals to purges and the Great Famine of 1932–3 (called the *Holodomor* in Ukrainian).

Though Stalin's projected collectivization was never completed, it damaged human–nature communities the most. The harm was especially great during the first five years – called the 'utopian fantasy', the 'new world', or 'gigantomania' – when whole regions were redrawn on maps and hundreds of villages were joined together in massive state-owned collective farms (Fitzpatrick 1994: 105–6). The stages that followed entailed various programmes that did not require peasants to relinquish all their property to the state. In the *toz*, peasants kept their own land but shared machinery with others; in the commune, people gave up their private property altogether. In between was the *kolhoz* or collective farm, on which peasants from two or three villages kept some private plots of land but handed the bulk of property over to the *kolhoz*, which was managed by an elected committee under a chairman appointed by local representatives of the party (Lane 1978: 59).

Transplantation of Soviet Russia's collectivization model varied from country to country in the Soviet bloc and differed significantly in its pace. After Sovietization had essentially been completed by 1948, and communist governments predominated in Eastern Europe's states (Naimark 2010: 183–4), collectivization took place, but was not as extensive as in the Soviet Union and the non-Russian Soviet republics of Ukraine and Belarus. In Hungary, collectivization was implemented by force beginning in 1949, and large state farms were established in place of private holdings (Balassa 1960: 36–7). Poland, however, resisted collectivization due to a strong attachment to traditional rural culture (Jarosz 2014: 120), heritable structures of land ownership, rural–urban migration (especially in the 1950s) and the authority of the Catholic Church (Jarosz 2014: 130–2). Despite the differences among these countries, the Stalinist framework of 'modernization' stabilized Soviet communism until 1991, when it collapsed in Eastern Europe and state-owned farms were finally liquidated.

It is necessary to ask how peasants contributed to environmentalist resistance against the communist government, which imposed environmentally damaging modernization on villages and caused massive destruction to their natural surroundings. Although direct acts of political rebellion did take place during the long process of collectivization, I understand peasants' environmentalist resistance in a specific sense – as a resonance in culture and, primarily, in literature, which explores interconnections among humans, animals and soil tired of collectivization. The tired village is the voice of a tired land – in fact, it is a polyphony of voices, rejected at the time and here reconsidered in the framework of cultural memory.

2

Fatigue: Platonov's *Pit* and the Stalinocene

The story of the weary village takes place in a critical moment for cultural memory: the darkest period of Stalinism. The village is exhausted by a radical crisis in human–nature relations, but at the same time the figure of the tired village provides a way to go beyond the anthropocentric framework usually used to study peasant culture and recognize narratives that enable readers to hear the 'voices' or 'soundings' of non-human actors. Despite the tragic context, which such stories reveal to an unnerving degree, this period of Stalinism was a critical moment for representing agricultural environments, as the example of the Soviet countryside and its periphery shows.

In this chapter, written in the shadow of Andrey Platonov's *The Foundation Pit*, I refer to Russia during the worst period of Stalinism, a time of intense collectivization, hyperindustrialization, purges and hunger, when life itself was rendered absurd. Historians have called the first Five-Year Plan, from 1928 to 1932, 'The Great Rupture' (Heller and Nekrich 1986: 232) – something like the Great Acceleration in the Stalin era. This period was destructive not only for humans, but for all beings. We might expect that such conditions could only deprive people of agency and starve their imaginations of creative activity, but through literature people also resisted despite the shocking scale of Stalin's crimes. As in the case of collective trauma experience after the Holocaust in Jewish cultural memory, what Stalin's Five-Year Plan enacted was the first genocide against the state's own population in times of peace – the genocide of the peasants (Heller and Nekrich 1986: 236). The peasants who did not support the new order, so-called *kulaks*, were stripped of their humanity, as the Nazis dehumanized Jews. The memory of these events created a continuous need to refer to, represent and reconstruct them.

Historically, the first Five-Year Plan was also the time when devastation of the natural environment increased most rapidly, through water pollution, unrestricted logging, soil contamination, destruction of landscapes, intensive development of mining and heavy industry, collectivization, and overambitious hydroengineering projects such as the construction of Belomorkanal (the White Sea Canal), which cost thousands of human deaths. Soviet conservationists' opposition to this canal was ultimately unsuccessful, and they were persecuted like other political enemies (Weiner 2000: 120). The first Five-Year Plan was not only about developing the economy, but also, as Maxim Gorky put it in his speech at the opening of Belomorkanal in 1933, about how 'Man, in transforming nature, transforms himself' (Weiner 2000: 119). Historians emphasize that this period of 'state-led, centrally planned gigantism' transformed the country into a superpower able to defeat Nazi Germany during the upcoming war, but it came at a high, irrevocable cost in terms of the damage done to people and nature (McNeill and Engelke 2014: 131), leaving psychological as well as ecological scars.

The years of Stalin's personal reign, according to historians, signify 'the most crucial period for understanding the Soviet Union' (Hosking 1985: 12). Stalin believed that he was creating something extraordinary (including overcoming nature). Thus he 'did not want to be reminded of his real name, Dzhugashvili, and immediately instituted the sonorous "Stalin" – from the word "steel" (stal)' (Sinyavsky in Hosking 1985: 82) to introduce the new steel era, which I call the Stalinocene. This term is geographically and historically specific in relation to the discussion around the popular in humanities 'Anthropocene'. The Anthropocene locates a new geological epoch of human decisive influence on environmental degradation and global warming in different time periods, beginning with the Western world, capitalist, Industrial Revolution (see Garrard, Handwerk and Wilke 2014), while historians of the Soviet Empire claim that, in contrast to the tsarist and Bolshevik governments, the most disruptive period for environmental history in this part of the world happened during the Stalin leadership (Josephson et al. 2013: 71).

The essence of the Stalin era was captured in a newsreel, *5 marca 1953 roku zmarł Józef Stalin* (On March the 5th, 1953, Joseph Stalin died) (episode 11–12/1953), released just after the death of Stalin. This particular newsreel is a special edition made in Poland as a tribute to 'the good father of nations', although similar propaganda newsreels were released in all Soviet bloc countries. The newsreel represents industrial development through images of massive plant construction sites, gigantic substations, industrial fumes, smog in Moscow, smelters, traffic jams and car pollution – images that, for us, evoke environmental catastrophe and recall the cityscapes of early Western capitalism during the accelerated industrialization of the second half of the nineteenth century (London, Manchester, New York and Chicago looked the same less than a hundred years before). The voiceover of this propaganda documentary, however, proudly describes what Stalin and his epoch accomplished:

> Under the command of Stalin, the formerly backward country has become a mighty industrial-*kolhoz* power. He raised sturdy generations of Soviet people, builders of a new life. Great Stalin, the teacher, the chief, the father. Stalin – the continuator of the immortal Lenin's work. Moscow – the city of Stalin, hope for the world. In Stalin's native land, the long-time human dream of peace and happiness has materialized. Stalin leads the Soviet nations to communism. Stalin ordered the rivers to turn back their courses and changed the deserts into fertile lands. Stalin stopped the damaging winds and transformed the landscape of the Earth. Stalin is mankind's tomorrow. Stalin is the name of the epoch.
>
> my translation

In fact, for Stalin nature was as much an enemy as his political opponents. It was necessary to tame nature and adapt it to function in the new communist state. Nature's capricious, irrational character was the epitome of capitalism, which had to be subjected to the rational and coherent policy of the Communist Party (Josephson et al. 2013: 74). In his 'theoretical writings', for example, Stalin speaks about combating droughts (1973: 263) and floods (446–7). Thus, unlimited industrialization and collectivization involved not only factory and farm production, but also a broader plan for exploiting nature, which was 'a participant in the communist project' (Bruno 2016: 7). While Engels, in *Dialectics of Nature*, praises evolution and Darwin's contribution to materialism (Stalin 1973: 302–3), Stalin understands progress in a revolutionary way, as a transition from Engels's biological perspective to total industrialization, which was to replace the formerly backwards and quintessentially Russian peasant society.

Stalin's vision of the economy rejected the Russian cultural landscape and its history. By eradicating peasants, Stalin eradicated part of the Russian soul and their connection with the soil. He was sure that, under the socialist system, Russia would no longer be divided into industrial and agrarian regions, since all lands were 'more or less industrial' (262) and 'the antithesis between town and country is disappearing' (266). He explained further that 'the countryside itself now has its own industry, in the shape of the machine and tractor stations, repair shops, all sorts of industrial undertakings of the collective farms, small electric power stations, etc. The cultural gulf between town and country is being bridged' (266). No matter how emancipatory this rural transformation appeared – and it certainly did advance gender equality and educational progress – the environment benefited more from the old divide between city and country.

So how did the collectivization of villages proceed? *Combinates* and *kolhozes* had replaced villages before the Soviet bloc countries came under Stalin's leadership in 1928, and became one of the many spaces where Eastern Europeans were oppressed. These collective farms were widely established in Soviet bloc countries via the state's land confiscation and created a new type of peasant – a labourer, a forerunner of the utopian worker – who was incentivized to work harder and praised for being a shock worker. In propaganda materials, there was nothing left of the peasant who represented the melancholic Russian soul, which had been so frequently painted by nineteenth-century artists, as in Ivan Kramskoy's portraits, and depicted by prominent writers, as in Ivan Turgenev's *The Hunting Sketches*.

In literature, one of the bitterest images of proletarian labour and the *kolhoz* from the period of the first Five-Year Plan in the Stalinist Soviet Union was created by Andrey Platonov in his counter-propagandist narrative *The Foundation Pit*, which superficially resembled the officially accepted genre of the so-called 'production novel'. This genre was part of the Socialist Realist canon and represented a collective rather than an individual hero confronting challenges in a naively moralistic way. The conventional plot involved 'the fulfilment of an industrial assignment, be it the construction of a new plant, the establishment of a Stakhanovite record [from the name of a coal miner, Aleksei Grigorievich Stakhanov, who massively exceeded extraction targets] or the implementation of some daring technological invention. In all cases, these goals require the symbolic or real death of central characters as a sacrificial offering' (Kahn et al. 2018: 676). Platonov built on this conventional narrative – and, in fact, reversed it – via two intertwined settings of a construction site and a collectivized village, formerly divided and now to be connected. Apparently, he wrote this tale in a few months between 1929 and 1930, but it did not appear in print in Russia until 1987, long after the author's death in 1951. The Soviet-style rural economy had regional and state variations that translated into different environmental consequences and, as a result, different cultural responses to this period: for example, Ukrainian and Czechoslovakian farming were completely collectivized, while family farming remained predominant in Poland apart from the state agricultural farms (so-called *PGRs*). Despite this variation, Platonov's *The Foundation Pit* can probably be regarded as the first literary text to respond to collectivization during its first and most exploitative wave. It is also the first anti-Soviet testimony that underscores not only the vulnerability of the people, but also that of nature and the land, written with full awareness as a response to this dark Stalinist stage of economic revolution.

Platonov himself was born in 1899 in Voronezh, close to the border with Ukraine. The family lived in a small settlement bordered on one side by Voronezh and on the other by railway tracks, beyond which lay only wheat fields, forests and the high steppe (Seifrid 1992: 3). Platonov spent his earliest years in a kind of half-peasant, half-proletarian world at the intersection of the natural and the mechanical, which fascinated him early on (Seifrid 1992: 3) – he himself was involved in hydroengineering and electrification projects – but that later became a source of his critical, pessimistic 'dark' writing. 'From this origin on the "margin"', as Platonov's biographer Thomas Seifrid observes, 'between two worlds – rural and industrial, old and new, natural and man-made, traditional and revolutionary – derive many of the contradictions that characterize this writer and his work' (1992: 3). He devoted himself totally to writing but did not praise socialist politics; he instead produced something like a satire of collectivization, *Vprok – Bednyatskaya khoronika* ('For Future Use. A Poorpeasant Chronicle', 1930), written from the point of a simple peasant, and was subsequently banned from publishing (Seifrid 2009: 25). He suffered a damaging experience when his son was taken to a labour camp in the far north; a photograph from 1938 shows how Platonov aged almost overnight (see Figure 1). With the aid of the more

Figure 1 Andrey Platonov in 1938. ©Maria Andreevna Platonova.

influential writer Mikhail Sholokhov, the boy was released and returned home in the spring of 1940, but he was fatally ill with tuberculosis and infected Platonov (Seifrid 1992: 13). After the death of his son in 1943, Platonov managed not only to survive – largely through the efforts of his wife – but also to continue writing. He eventually died of tuberculosis in January 1951 and was buried next to his son in Vagonskaya Cemetery. In addition to *The Foundation Pit*, he devoted many short stories to the problems of socialist industrialization and the struggle for socialist transformation of the countryside.

During this early period of Platonov's engagement in building the socialist state under the Soviet communist authorities, he was a utopian practitioner. He left his literary writing for the first time in 1924 to work on engineering projects, such as land reclamation in his native Voronezh region (McKenzie 2015: 67). From the beginning, these and other projects of the early Soviet state were overambitious and heedless of natural environmental capacity. Soviet engineers like Platonov treated rivers, lakes and wetlands as strange machinery for building a utopian, self-sufficient system that would prevent hunger. Of course, such utopian pursuit of technological progress belonged not only to the early Soviet period but was as central to capitalism as communism around the world (e.g. building hydroelectric dams on the upper Columbia River or coal-fired power plants in Arizona's Four Corners, which displaced many native people and brought high environmental costs to stimulate urban consumerism during the New Deal reforms, Klingle 2014: 486). The Stalinocene, however, raises a set of questions about communism versus capitalism. Those who advocate the term 'Capitalocene' argue that capitalism in particular is responsible for climate change and environmental damage; communism, however, especially in the Stalin era, accelerated production of output to an extent that left far behind other countries, such as the United States or Great Britain, when we compare the years between 1928 and 1950 (Josephson et al. 2013: 86). It was not, though, aimed at raising income or stimulating consumerism, but at praising success over the capitalist world in communist propaganda, while starving the people, exploiting them in slave labour camps (gulags), and irretrievably damaging the environment.

According to Viktor Shklovsky, in the early Soviet period Platonov was part of this utopian progressivism (McKenzie 2015: 67), but he naively believed in the possibility of its universally profitable application: 'that technology ... might with one grand and magical stroke transform the countryside – irrigation schemes, electrification, railways' (Jordan 1973: 53). Especially in *Chevengur*, Platonov expresses the view that socialism or communism (he uses both terms interchangeably) 'is about getting water onto the steppes', which would glisten across the whole universe (McKenzie 2015: 71–2). This utilitarian and aesthetic perception of nature as a cosmic element was characteristic of the environmental culture of the early Soviet era, despite the dominant understanding of nature as an obstacle to be overcome, as Stalin thought. *The Foundation Pit* registers the depth of Platonov's disillusionment with this ideology and the social-ecological project it sustained.

Seifrid observes that, unlike other writers in the 1920s and 1930s, Platonov 'had direct, physical experience of the construction of socialism in the Russian countryside' (2009: 11). In particular, in the Voronezh region, much infrastructure to prevent future droughts was built quickly. We find in the documents that, during Platonov's work there from 1923 to 1926, workers 'excavated 763 ponds, dug 331 wells, drained 7600 *dessiatins* [1 *dessiatin* = 10,925 square metres] of land and irrigated 30, built bridges and dams and installed three electrical stations' (Seifrid 2009: 11). When we consider later hyper-industrial projects, such as building Bielomorkanal and transforming Magnitogorsk into a gigantic metallurgic city, these pre-Stalin years, along with his reign, seem very much to belong to the Anthropocene, especially because they involve humans subjugating geology and digging – one of the main activities of *The Foundation Pit*'s protagonists. The geological aspects of reality permeate into the workers' and peasants' community experience and into their future relation with the soil within this unknown construction project, as in Elżbieta Połońska's 1923 Soviet poem *W pętli* ('In the noose') about building a 'gord', or medieval Slavic fort:

> The whip whistled. They were digging a trench.
> They were laying a stone. They were vaulting the ceilings.
> Slaves were building this gord in the swamps
> In the country of subarctic utopias.
>
> 1979: 119, my translation

'The country of subarctic utopia' – the utopian effort to build a home for humanity in the new communist epoch – is a critical benchmark for Platonov, since he links it with the environmental impact of the Stalinist industrial programme. In earlier works, he directly addresses ecological catastrophe due to humankind's war on nature and efforts to erect a socialist home for humanity 'out of the clumsy, formless, and cruel earth' or to 'transform the geography of the planet to human benefit' (Seifrid 2009: 143). In a repeated motif, constructing this abstract, idealistic home for all people, whether with human hands or machines, results in the destruction of the earth. Such a utopian project also shapes the environmental context of *The Foundation Pit*.

Voshchev, one of the main heroes with whom *The Foundation Pit* begins, has been dismissed from the machine plant due to 'weakening strength in him and thoughtfulness' that slowed down the pace of his work (Platonov 2010: 1). Maybe because of his tiredness, or maybe because of his reflective, melancholy nature, from the beginning he stands out from and evades the rules of the socialist industrial regime's hyper-efficient machinery. He resembles the philosophical, Coleridgean figure who appears in many Platonov's texts – 'The Wanderer' who searches for truth (Seifrid 2009: 8) and, according to the author's notes, someone who 'feels everything' (Seifrid 2009: 140). In the poem 'The Wanderer', Platonov says that this figure is the one who *will see and understand*, while for the rest of us nature is an empty, deadly space, devoid of meaning and feelings, stripped of variety: 'In the field is neither a wind, nor a cry, / Nor a lonely white willow' (Platonov 1993: 348).

Voshchev is a contemplative and oversensitive man who has the audacity to reflect on 'a plan of shared, general life' (Platonov 2010: 3), who is slow and vulnerable to the acceleration of work in the Stalinocene, who is not afraid for his own life. He represents what I call 'weak subjectivity', a melancholic and passive figure of a man who substitutes his own individualistic needs with relational approach to the surrounding world. While he is certainly lonely, he perceives the outside reality as meaningful and speaking directly to him: a tree that 'sway[s] from adversity' (Platonov 2010: 1–2), 'the questioning sky' shining above him 'with the tormenting power of stars' (2), the dog with 'a weak voice of doubt' (3). All these voices in the opening of the book show how sad – how unbelievably bleak – Voshchev's world is. This sadness is not about his life but the complete absence of happiness in the mechanized and soulless world where 'without thought, people act senselessly' (4). He watches from the edge of the city, where working people seem to him to carry too heavy a burden: 'Man puts up a building – and falls apart himself. Who'll be left to live then?' (9).

In Voshchev's eyes, this technological anti-life is suddenly pouring out of the human world, as in the land of Hans Christian Andersen's Ice Queen, causing sadness and death all around, although at first this tragedy is only delicately indicated. An ordinary fallen leaf is a 'dead' leaf, another object of 'unhappiness and obscurity' (5); the passing procession of young pioneers have 'frail', 'hardening' bodies, and will never fit into the landscape of fields on which 'dead horses of social warfare were lying' (6), and 'only birds could sing the sorrow of this great substance, since they flew up above and life was easier for them' (9). In the end, this sadness will turn into a total deadness and reveal a multitude of corpses and animal carcasses as in a battlefield after a war. It is as if the memory of the Stalinocene could not cope with all its cruelties but froze them in the novel's apocalyptic landscapes and in its description of an ecological trauma.

From the very beginning of *The Foundation Pit*, humanity embodies death – first when a man holding an agricultural scythe enters a deserted, 'empty wilderness' and begins 'to mow grassy thickets that have grown there since time immemorial' (9). Anxiety fills Voshchev as he digs and destroys 'thousands of rootless, grass blades, and little soil shelters of diligent creatures' (13), and when he only smells grass 'that had died and the dampness of bared places, making more palpable the general sorrow of life and the vain melancholy of meaninglessness' (12). But this digging is only the first stage of destruction. Its next layers are revealed by a Soviet engineer, the paradigmatic incarnation of the Stalinocene:

> Amid the wilderness stood an engineer – not an old man, but gray from the calculation of nature. He pictured the whole world as a dead body, judging it by those parts of it that he had already converted into structures: the world had always yielded to his attentive, imagining mind that was limited only by an awareness of the inertness of nature; if materials always gave in to precision and patience, then it must be deserted and dead. But amid all this mournful substance, man was alive and of worth.
>
> 12

Digging means 'working in cramped narrows of dreary clay', 'pulveriz[ing] the compressed layers of rock lower down', 'annulling nature's old order', and breaching 'the age-old ground' (13), as if Voshchev had a guilty conscience. While one of his comrades in work exclaims that 'the earth needs the touch of iron or it lies there like some fool of a woman. It's sad!' (13), Voshchev tries to justify their activity: 'Maybe nature will show us something down there' (16). He himself looks for some deeper way of understanding life spatially and materially – a comprehension that he may find in the pit, the universal foundation, the whole world's building block hidden in the ground. We know already that this foundation will not be socialism.

On hunger rations, workers carry on with the remnants of their strength, and Platonov does not spare us descriptions of their fatigue and weakness. His biographer emphasizes that Platonov himself witnessed the cost of the collectivization campaign in the tragedy of famine (Seifrid 2009: 97–8). Historians speak about the ultimate annihilation of the traditional peasantry and Stalin's goal of transforming them into 'a class of industrialized rural workers' (Seifrid 2009: 98). Nature accompanies this exertion and weariness in the tale, not only figuring in the visions of Voshchev's tired and hungry body, but also participating in the phantasm of transformation. For example, a dead bird prompts Voshchev to recommit himself to the empty hope that this stage is like a state of emergency that justifies continuing the excavation:

> The sun was still high, and birds were singing plaintively in the illuminated air, not in triumph and celebration but searching for nourishment in space. Over bent, digging people swallows were hurtling low; tiredness stilled their wings, and beneath their down and feathers was the sweat of need; they had been flying since first light, ceaselessly tormenting themselves to fill the stomachs of their chicks and mates. Once Voshchev picked up a bird that had died in an instant in midair and fallen to earth; the bird was all in sweat, and when Voshchev plucked it, so as to see its body, what remained in his hands was a scant, sad creature that had perished from the exhaustion of its labor. And now Voshchev did not spare himself in annihilation of the close-knit earth. Here the building would stand; in it people would be stored away from adversity, and they would throw crumbs out of the windows to the birds living outside.
>
> Platonov 2010: 17

Voshchev, however, quickly loses his naive, transitory expectation of participating in something meaningful, as 'the grief of the general condition was beginning again to torment Voshchev; sometimes he sensed the whole external life just as he sensed his own innards' (23). The construction of the house is a utopian project, a project about which workers know little except for what the engineer Prushevsky has told them about it being a monumental, all-proletarian house in which people will no longer live 'fencing themselves off into households' (19). The novel, therefore, implies that this structure will literally house the masses. At the same time, the engineer is absorbed by a dream: 'far away a nighttime factory construction site was shining with electricity, but Prushevsky knew that there was nothing there except dead building material and tired unthinking people' (19). He lays out further absurd plans that 'after ten or twenty years, another engineer would construct a tower in the middle of the world, and the laborers of the entire terrestrial globe would be settled there for a happy eternity' (19). But now they have to dig in the gully, and others 'had yet to join in the task of construction' (43).

In the absurd reality of the construction site, the diggers 'acquire the meaning of mass life from the loudspeaker' (50). At one point they are informed that their 'task is to mobilize the stinging nettle onto the Front of Socialist Construction!' because 'beyond our frontiers the stinging nettle is nothing other than an object of crying need!' (50). Everything that exists has to be subordinated to a total project of transforming the human and non-human world, even plants commonly known as weeds. Everything alive and useless now must become part of the socialist plan. Nothing can be left alone; nothing can just exist outside the communist society's purpose – it has to be put to work somehow. Another time, the task 'is to cut off the tails and manes of horses!' to get tractors or pick up snow from collectivized fields (51). When the loudspeaker's stream of absurd challenges continues, making the workers despair, one of Voshchev's comrades challenges him to a 'socialist competition for the highest happiness of mood' (51). Not only does Platonov ironize this challenge, but he also mocks the specific propagandistic language that was used to speed up people's work, even if a faster pace was beyond their capabilities. Another ironic figure is the 'forsaken peasant' who stays with the diggers but does not fit into the new order: 'what kept appearing to his yearning mind was a village in the rye, the wind blowing above the village as it quietly turned the sails of a wooden mill and ground the flour for his peaceful daily bread' (57). Now, even his tears have to be quick, without delay, but he echoes the romanticized figure of the peasant in Eastern Europe, who lost his rural world and has been rejected by the communist system. He tries to explain his sorrow, apologizing for his backwardness in the new reality (57).

The *kolhoz* located nearby is a newly transformed village. Voshchev, who wanders there following the poor peasants, is asked by an activist in charge of new life in the collective farm to check the hens. He does not find 'an egg under a single bird', and the activist's assistants comment that it might be a sign that the fowl are 'really prokulak' (79) – that is, standing with some individual peasants who did not want to collectivize. Although the hens do not have a cock in their flock because he was eaten by starving men, they are suspected of being 'a base for another kulak superstructure' (79). The absurdity of collectivization penetrates deeper into the farm animals' world. Voshchev notices, with horror, 'socialized' horses that, without human supervision, live equally in herds and obey the new organization of life like companionate robots:

> without the labor of man a gate to their right had opened and out onto the street began to emerge calm horses. At a steady pace, without lowering their heads to the growing nourishment on the earth, the horses walked in a close-packed mass along the street and down into a gully that contained water. After drinking their norm, the horses entered the water and stood in it some time for their own cleanliness; they then climbed up onto the dry land of the bank and started back, not losing their formation or compact solidarity. But when they came to the first homes, the horses dispersed – one horse stopped beside a thatched roof and began to pull straw from it, another bent down to pick up in her jaws some

residual wisps of meagre hay, while the more sullen horses entered the farmsteads, took a sheaf each from dear, familiar places, and carried these sheaves out onto the street in their mouths. Each animal took a share of nourishment proportionate to its strength and carefully carried it in the direction of the gate from which all the horses had first emerged. The first horses to arrive stopped outside the shared gate and waited for all the remaining equine mass; only when all had assembled in common did the leading horse push the gate wide open with her head and the entire equine formation move away into the yard with the fodder. There the horses opened their mouths, the nourishment fell from them into a single heap in the middle – and the collectivized livestock then stood around and slowly began to eat, having made its peace in an organized manner and without the attention of man.

88–9

Even if this scene that Platonov so evocatively describes is only a figment of Voshchev's imagination, or hallucination, the non-human collectivization that takes place before his eyes is absolutely unreal. The peaceful and harmonious collectivization of the domesticated horses confirms that the project of Soviet socialism, especially in the Stalinist period, is a totalitarian enterprise that involves all – humans as well as non-humans. Because this description of a commune of 'calm' horses – a perfect 'formation' of 'compact solidarity' that is able to follow socialistic rules, such as eating and drinking according to certain 'norms' – is an exceptional fragment in the whole book, readers are left to conclude that humans cannot behave in such an orderly way. Perhaps this passage was intended to serve as a 'satire on the "voluntary" nature of collectivization' or to express the utopia of 'collectivity extending throughout nature' (Seifrid 2009: 124).

When the imagination no longer helps and the collective wins, the only question is: 'Is it really sorrow inside the whole world – and only in ourselves that there's a five-year plan?' (Platonov 2010: 33). Let us not forget that we are dealing with the first stage of Stalin's economic project, the cruellest period for humans and environment. The trauma caused by Stalin's 'revolution' is so totalizing that the human world becomes blurred with that of non-humans. 'The eternally passive acquiescence of nature' (Platonov 2010: 93) is vulnerable to the communist revolution, but, for those who resist collectivization without knowing that they are doing so, this revolution brings death. In this deeply ontological transformation of the world, everything that is non-collectivized is marked by death – even a sparrow in the Orthodox Church who 'looked in silence at a man, evidently meaning to die soon in the darkness of autumn' (93). Platonov, though, extricates those voices lost to communism and preserves them in the traumatized memory of human and non-human acts of witness.

Some peasants, knowing their fate, tried to oppose this reversal of reality. They would do anything, even eat or starve their own animals in order not to give them to the *kolhoz*:

> The residual, non-collectivised horses slept sadly in their stalls, tethered so firmly in order that they should never fall, since some horses already stood dead on their feet; in anticipation of the collective farm the less impecunious peasants had kept their horses without nourishment, so that they would enter social ownership only with their own bodies and not lead their animals after them into sorrow.
>
> 101

'The peasants are mostly nameless and faceless in *The Foundation Pit*, as they were to the Stalinists, but they are not voiceless,' writes McKenzie (2015: 87). Collectivization changed the peasants' connection to the soil, their animals and their own human nature. Therefore, it is a cruel act when they starve or slaughter their own livestock, but also one that the peasants consider anti-communist and politically meaningful, albeit not for the animals. This chapter of non-human history during Soviet collectivization has not yet been written, but *The Foundation Pit* opens the discussion by recognizing animals as its victims as well. This history can be traced through the broken bonds between peasants and their animals and interrupted a long period of mutual attachment and dependence.

Platonov quotes a rebellious peasant tenderly speaking to his mare, trying to comfort her while she is dying: he 'embraced the horse's neck, and stood there in his orphanhood, smelling in memory the mare's sweat, as when they were plowing' (101). For him, the farm world is ending in apocalypse, full of naturalistic images of starved humans and animals. The owner of the mare, as if in a trance, does not notice the dogs who begin tearing the meat from her legs and continues his monologue. The pain keeps her alive even longer – she cannot die. In this world of unbearable distress, the mare's 'life was just sinking into further places of poverty, breaking down into smaller and smaller parts, yet still not managing to exhaust itself' (102). The farm's reality is falling apart instead of sinking into nothingness and bringing relief. When peasants who do not accept the new order eat their animals, the ontological destruction of the old village and material consumption in the act of rebellion intertwine – and the human–animal relationship is completely destroyed: 'Some calculating peasants had long ago swollen up from meaty food and were now walking heavily, like moving barns; others were vomiting continually, but they were unable to part with their cattle and so they destroyed it down to the bone, not expecting the benefit of stomach' (102).

Basically, in *The Foundation Pit*, the world of people and nature either dies – its language does not distinguish the deaths of humans from those of non-humans – or is subjected to the newspeak ideology of collectivization. Aesthetics collide with propaganda, especially when Platonov poeticizes the demise of the village and its multispecies inhabitants: 'Cows and horses were lying in these yards, their carcasses gaping and rotting – and the heat of life accumulated during long years beneath the sun was still seeping out from them into the air, into the shared wintry space' (109). While animals who are called 'working class', like a tamed bear on the farm (109) or the 'socialized' horses, are part of the broader 'acquiescence of nature' (93), such compliance is expressed only in the contaminated language of the characters' delusions. The characters see through the lens of propaganda expressions that are so versatile they cannot escape them: even when they look at an empty, uninhabited space, they assign phantasmic categories to it. The anti-reality of the *kolhoz*, like the whole of this absurd project of industrialization and socialization, generates new corpses and death, breaking old ties, including those between people and animals. The unreality imposed by communist propaganda language leads to the real destruction of the village. Platonov expressed its consequences in a painful imaginarium that extends human suffering into the non-human world and creates an image of ecological trauma that is distinct in Eastern European literature.

Platonov is merciless to his readers, and near the end he includes more naturalistic fragments, such as people dying in sequence, flies feeding on a sheep's body, and descriptions of how warmth coming from slaughtered animals disrupted the seasons (110–11). Reading this text could be compared to listening to a musical symphony in which a rising and increasingly loud stream of sounds leads up to a culminating moment. The last grave-pit is prepared for a little girl who had been adopted by workers after her mother died.

The peasants crying for 'dear' and 'familiar' carcasses (102) belong to one of the most desperate fragments of the book – one that reveals the real, final pit into which both people and animals fall, and from the bottom of which no more life emerges. The pit widens semantically to reveal a frightening reality that we could not reproduce without Platonov's narrative, even though, in the novel, the reality of destruction and mass death disturbingly merges with the unreal language of propaganda.

By way of comparison, some quite informative non-fiction memoirs written by witnesses of Stalinist collectivization, such as Fedor Belov's *The History of a Soviet Collective Farm* (1998 [1956]), give accounts that only confirm and fill some gaps in the unimaginable reality Platonov depicts. Belov, for example, corroborates the sad atmosphere that followed dekulakization, as girls no longer sang in the evening and instead dogs howled on the ruined farms (1998: 6). Then he recounts details from the period of the Great Famine in Ukraine (1932–3), which he calls 'the most terrible and destructive that the Ukrainian people have ever experienced' (12). Some of his recollections evoke the context of *The Foundation Pit* and underline how realistic this fiction is:

> The peasants ate dogs, horses, rotten potatoes, the bark of trees, grass – anything they could find. Incidents of cannibalism were not uncommon. The people were like wild beasts, ready to devour one another. And no matter what they did, they went on dying, dying, dying.
>
> They died singly and in families. They died everywhere – in the yards, on streetcars, and on trains. There was no one to bury these victims of the Stalinist famine.
>
> Belov 1998: 12

Belov also describes the paradox of the centrally managed economy in the *kolhoz*: it creates obstacles to raising crops and sowing beets. The peasants must wait for central orders despite their agronomic knowledge about how to grow food (1998: 146–7). Historians provide similar accounts of the horrific conditions during collectivization. Peasant resistance, such as the starvation of livestock described in *The Foundation Pit*, took many forms, including women's riots (*bab'i bunty*), and the theft and destruction of collective farm property. Perhaps most widespread was an intentionally slow pace in carrying out directives from the *kolhoz* administration (Fitzpatrick 1994: 64–6). The tremendous loss of livestock through slaughter (Josephson et al. 2013: 74), inadequate fodder and simple neglect made it virtually impossible for *kolhozes* to fulfil their procurement quotas for meat and dairy products, which led to the famine of 1932–3 (Fitzpatrick 1994: 70–4). Heller and Nekrich (1986), as well as Fitzpatrick, use the term 'passive resistance' to characterize peasants' obstruction of *kolhozes*, such as by refusing to sow (62), through apathy (Fitzpatrick 1994: 66–7) and by destroying their livestock in protest (Heller and Nekrich 1986: 237; Fitzpatrick 1994: 64). While their decision to resist was fully understandable, for animals it resulted in decimation: between 1928 and 1932, the number of horses dropped from 33.5 to 19.6 million; cattle from 70.5 to 40.7 million; pigs from 26 to

11.6 million; and sheep and goats from 146 to 52.1 million (Heller and Nekrich 1986: 237; similar figures are given by Josephson et al. 2013: 97). Fitzpatrick even writes that 'the slaughter was a pure protest, accompanied by wild days of feasting as a kind of farewell to the old life' (1994: 66). The terrible loss of horses contributed to famine and production problems, and was also caused by the state's impossible plan for tractors and combine harvesters to supersede the animals (Fitzpatrick 1994: 136–7). In the 1930s, horses continued to be the main means of transport and draft power for ploughing and other agricultural tasks, but there were too few of them and they were collectivized by the state (Fitzpatrick 1994: 137), which caused many local political tensions (Fitzpatrick 1994: 138–9).

Platonov is perhaps the first writer in Soviet literature who grasps the tragedy of collectivization so well when he asks about the limits of these tragic and immobilizing reforms, promoted by Stalinist propaganda, that affected the human and non-human world, culture as well as nature. The writer personally experienced the communist project's failure to heal the human world from poverty and hunger by overusing natural resources. While, at the same time, he prevented the silencing of nature in cultural memory of the Stalinocene. Against 'the face of an unresponsive nature', as Marion Jordan wrote about Platonov's other work – *For Future Use* (1973: 27) – *The Foundation Pit* enacts an intensive search for comradeship with the nature silenced by the Soviet utopian project and entails the torture of witnessing the tragic fate of collectivized villages, humans and animals. The novel shows how communism was built against nature – it was a catastrophic project that perceived 'nature without ecology' (McKenzie 2015: 82).

Voshchev's sack

Despite being tired and lost in this Stalinist reality of the first Five-Year Plan, Voshchev carries a sack in which he puts 'all kinds of objects of unhappiness and obscurity' (Platonov 2010: 5), 'every kind of obscurity for memory and *vengeance*' (9, my italics), and in which things can exist 'as facts of the melancholy of each living breath' (44). The objects in the sack can be transitory, like leaves, or nearly eternal, like pebbles found below the ground 'in the midst of clay, in an accumulation of darkness' (23). Gathering this collection is the only act that comes from Voshchev's own will, against the despair and hopelessness of all existence. McKenzie interprets these collected objects as 'the worn traces of human life and labor' (2015: 89), but they are also signs of a world uncontaminated with propaganda and lost to that totally intrusive stage of human presence on the earth called communism. These common objects that Voshchev finds may provide evidence of the human merger with the non-human, the environmental remnants for which nobody will be left to testify. However, hidden in the sack, they are rendered useless for the socialist project and somehow saved, lodged deep in the memory of this traumatic period.

Platonov's works are generally interpreted as pessimistic in presenting 'the impossibility of designing a *dynamic equilibrium* of human and natural processes', so that his writing returns 'to the residues that can't be recycled' (McKenzie 2015: 66). In other words, Platonov, even though he tried, was not able to enfold all of the natural world into the communist project represented in his own literary work – as if he knew that this totalitarian machine for producing new meaning is porous and leaks. However, all these leftovers collected in Voshchev's sack have meaning because they can be reused or recycled in fiction, leaving outside the unsuccessful Soviet project of transforming and mobilizing all of the natural world. In effect, this reflects his original expression of his final disappointment in communism, since, before *The Foundation Pit*, he held the idealistic but also radical view that everything is connected in some kind of socialistic bond of comradeship. McKenzie quotes Platonov's 1922 letter to his publisher, in which he argues: 'There is a kind of link, some kinship, among burdocks and beggars, singing in the fields, electricity, a locomotive and its whistle, and earthquakes – there is the same birthmark on all of them and on some other things too … Growing grass and working steam engines take the same kind of mechanics' (2015: 67). McKenzie calls these words 'a strikingly poignant … account of everyday proletarian life, among rocks, animals, and plants – as comrades' (67). But in the same letter, Platonov also proves himself a true son of the rapidly industrializing society: 'I believed then that everything is man-made and nothing comes by itself; for a long time I thought they made children somewhere at the factory instead of by mothers producing them from their wombs' (Yevtushenko 1971: 9).

Therefore, *The Foundation Pit* is an example of how contaminated language can be interwoven in a disturbing and vividly real, though blurred, anti-communist narrative. In fact, the tension between what is real and what is not is key to understanding this tale, since the communist terminology in which Platonov's language is soaked comes from actual discourse about the real Five-Year Plan. The poetic significance of many of the non-human actors involves their unrealistic status, which contrasts with the dying world of the collectivized village, since they also belong to nature and the surrounding landscape of the novel. However, they mirror this world in a distorted way – perhaps in Platonov's kind of magical realism – as if there were a parallel reality indicated by 'the music carried off by the wind, across the empty waste by the gully and into nature' (Platonov 2010: 2). The poetic elements of reality, though, are mainly seen through the solipsistic world of Voshchev. His visions appear during the starvation and extreme fatigue that he and the other human protagonists endure, but he also recognizes the corollary of their suffering in the natural world and finds a space in his imagination to preserve it – in his memory sack.

3

The Rural World is Gone: Peasants' Voices

This chapter aims to trace what has changed in the pastoral literary image of the traditional rural world during communism; and how the characteristics of the Eastern European cultural memory trope of a triad of relations among peasants, animals and soil have been affected by the Soviet project of transforming the village into an industrial unit of economy. How was this pastoral image redefined in cultural memory, and how did it function in rural literature resistant to the radical changes in the village during collectivization? However, we will find texts which used propagandistic language and followed official guidelines, and texts that passed muster with the authorities and were written to promote collectivization, while in the last chapter Platonov's *The Foundation Pit* was only published long after it was written because it was too critical of this 'contaminated' language to be published at the time.

Other writers analysed here represent anti-Soviet peasants' voices when they refer to the village. In an ecocritical sense, their voices anchor a community interdependent with nature and history. They show how a pastoral ideal of the village resisted environmentally destructive agrarian reforms, how it fell apart during successful collectivization (as in Platonov's *Foundation Pit*), and how it was rejected and poeticized. This rural literature can be reread in relation to environmental history of the village during the Soviet period. Because their texts are full of ecological concern, whether inscribed in alternative narratives of collectivization (Zabolotsky) or stories of the fading old rural world (Village Prose), their writing involves a new variant of the pastoral that belongs to peasants' environmental culture, which is preserved in the literary memory of Eastern Europe and described in this chapter.

Rural life was rooted in repetitive, cyclic and tiring work, and was imagined in pre-Soviet culture as ahistorical and microcosmic, belonging to a world that remained obscure until collectivization. After the Russian Revolution, some peasants were easily manipulated by communist authorities and became a social class whose 'emancipation' was an aim of the new, equal socialist order. Many others, however, responded critically to reforms that would destroy their way of life by developing a 'subaltern' peasants' voice (Fitzpatrick 1994: 4). It seems, from historical studies, that more peasants were against the new rural politics than for it, although those who advocated the new order quickly started to adopt the propaganda 'newspeak' – as George Orwell called it in *Nineteen Eighty-Four* – imposed by official media. They were highly motivated to express their opinions within the propagandistic framework, describing their own communist career paths and encouraging others to follow socialist models of devotion to developing the state's economy. According to propaganda materials, young people should try to become Stakhanovites (Fitzpatrick 1994: 10), i.e. 'normbuster[s]' (12), an ambition that spread through the countryside where young peasants became the rural motor of industrial acceleration. The aim of Stakhanovite propaganda was to 'encourage individual initiative in raising production that was born in industry and transferred to the countryside in the mid-1930s' (Fitzpatrick 1994: 12). Official propaganda newsreels (see Figure 2) presented images of villages where

Figure 2 'Let's fulfil and exceed the new five-year plan! More beets mean more sugar!' Propaganda poster by Vatolina Nina Nikolaevicha. 1946. ©Poster image provided by Poster Plakat.com.

'peasants were happy and not resentful of the Soviet regime; there was always an atmosphere of celebration; the sun always shone' (Fitzpatrick 1994: 16).

The Stalinist purges affected residents of the countryside, but their world was also destroyed by modernization and the intensification of agricultural production through such newly implemented practices as spraying toxic pesticides. The long-term use of DDT, for example, 'starved the soil of natural nutrients' (Feshbach and Friendly 1992: 54). Traditional sustainable farming techniques – such as 'shallow tilling, contour plowing, crop rotation, letting land lie fallow, fertilizing with organic compost and animal manure' (Feshbach and Friendly 1992: 58) – were replaced by mechanization (that is, use of tractors and combine harvesters), massive drainage and land reclamation. Scientific and technological progress were considered more important than the ecological capacity of the land and caused serious environmental damage, such as soil and water pollution.

Not only was the rural environment contaminated, but so was language. This *contaminated language* was used in all kinds of writings that more or less followed the guidelines of the authorities and promoted collectivization. The first contaminated texts related to the rural world emerged from the new state-sponsored subgenre of '*kolhoz* literature': peasants' diaries, short stories, novellas and novels in this genre were written from the beginning of collectivization in 1928 through the institutionalization of Socialist Realism in 1934 (Parthé 1992: 66). *Kolhoz* literature 'was charged with proclaiming and promoting the program of rapid, uncompromising rural transformation' (Parthé 1992: 66).

In parallel with the Union of Soviet Writers, which was established in Russia in 1932, post-war writers' organizations in communist countries propagated the same model of literature in all of Soviet Eastern Europe, until the Thaw period – the liberalization following Stalin's death in 1953. Rural themes were still exposed in all of post-war literature (Clark 1989), regardless of Stalin's era coming to an end. The so-called shock workers' propaganda texts, in particular, serve as historical documents of how extensive collectivization was carried out in villages and illustrate not only its environmental consequences but also its psychological ones. Authorities encouraged peasants to write diaries by holding competitions for writing in the style of Socialist Realism, which, according to communist officials, was the best way to convey the 'true' lives of individuals from the labouring masses in literature and represent their real conditions and their happy, fair and active communities. Moreover, diaries were promoted as 'eye-witnessing': they could reassure others that 'there is no reason to doubt the glorious future of the socialist agriculture and those who saw [it] turn into "new people"' (Mocanu 2011: 179; on competitions in Poland, see Wylegała 2017: 277–9). Despite the fact that such competitions helped many farmers develop literary careers, they were completely forgotten after the Soviet bloc's collapse because their work was artistically poor and saturated with propagandist slogans that had become devalued.

Another propaganda instrument was the newsreel. These short films of a few minutes each were narrated according to ideological needs and circulated across Soviet states. Now, the newsreel shows how language can be contaminated by the extensive exploitation of people and nature hidden behind the newspeak. Newsreels, especially from the post-war Stalinist period of 1946 to 1953, praised the accelerated pace of agricultural production to feed 'people', not 'landowners'. 'Everything must be developed, everything tilled and sown' (12/1946). Any obstacles in the path of this hyper-progressive agriculture, like the potato beetle infestation, are 'actively' eliminated by fighting back with pesticides (29/1948). The state aimed to improve livestock breeding in order to produce as much meat as bread, which was categorized as the development of 'breeding culture' (7/1949). A newsreel about a factory farm shows how dairymaids in one of the Ukrainian *kolhozes* use the milking machines to 'rationalize' production (26/1949). Only in a socialist economy, another newsreel claimed, could a 'gigantic plan for changing the climate and autonomizing meteorological conditions' be developed to transform the steppes and deserts into farmlands (41/1949). In *Fight for Fertility* (21/1951), 'backward' lands are destroyed to irrigate cotton crops: 'bulldozers are backfilling old canals and replanting old trees' and 'mighty machines' are carrying out a plan for 'more and more efficient work and a better life' for the 'new man'. In the *Days of the Grain* (9/1951), a peasants' commune 'undertook a call to accomplish the plan for storing the grain before the deadline'. Smallholders are shown turning over their 'substantial grain surplus' to the state, in contrast with so-called *kulaks*, the peasants who are 'hiding and destroying the grain' in order not to give it to the state. The narrator of *Harvest of Peace* (30/1951) describes how machines – the 'artillery of agriculture' – are helping people out. Peasants from a collective farm 'tackle July's tasks en masse, gripped by the example of the working class's sacrifice, they work faster and more efficiently'. The newsreel *Mechanized Haymaking* (38/1951) focuses on a scientific institute at a collective farm where workers use a variety of new machines, including one for laying haystacks. It concludes: 'the foremost agricultural economy of the USSR enters the period of full mechanization'. *Sowing for Peace* (21/1952) presents a *sovkhoz* [state-owned in contrast to *kovkhoz* – collective-owned, but in fact both were exploited by the state] type of farm, where planes are 'enriching the soil with artificial fertilizers' because 'the leading Soviet agriculture is fighting for new, even higher fertility rates'. *On the Fields of Kazakhstan* (44/1952) displays a giant sprinkler that Soviet scientists used on cotton crops: it sprays a 'life-giving rain directed by man' that waters twenty-two hectares of land within 24 hours. The sprinkler's long arms over the fields are described as 'wings of rich harvest, as if heralding a new era for man who has harnessed nature'. These examples highlight the priorities of the Soviet authorities: making rural lands more productive with science and machines, speeding up peasants' work and encouraging distrust of 'bad' peasants (the *kulaks*). The newsreels also illustrate the kind of new language consistently used to delineate these priorities.

The contaminated language of the communist period survived in younger generations' cultural memory and was rediscovered by post-communist authors, such as Wioletta Greg, who refers in her prose and poetry to her childhood on a Polish farm. In her short story, "The Little Paint Girl", she mentions, not without irony, how she was unintentionally immersed in Cold War propaganda: 'I had won a province-wide competition titled "Threats around your farm." I had painted a potato beetle climbing out of an empty Coca-Cola bottle.... The jury at the provincial level concluded that my drawing "portrayed, in a deeply metaphorical manner, the crusade of the imperialist beetle"' (2017: 31).

Literature of the 1960s and 1970s sharply criticized collectivization, but before then the critical voices of peasants or those advocating for them were rarely heard because of strict censorship and the repression of writers as well as other political dissidents. However, despite the difficulty of publishing 'bourgeois' and 'anti-Soviet' literature, some authors did address the 'contaminated' language of newspeak and destructive reforms in villages during the Stalinist regime. Such authors refracted these themes according to their responses to the regime's anti-ecological politics.

Nikolai Zabolotsky, a poet of peasant origins, responded to the fading, pre-collectivized rural world in the early Soviet period. Zabolotsky was victimized during Stalin's era of terror – he was arrested in 1938, and spent a few years in a gulag and in exile – and rehabilitated just after Stalin's death only to publish a significant collection of poetry (Yuzefpolskaya and Rueckert 2006: 290). His long, dialogic poem *Agriculture Triumphant* (Zabolotsky 1999: 61–84; in other translations, *The Triumph of Agriculture*), written between 1929 and 1930 and originally published in 1933 in the literary journal *Zvezda*, ambiguously portrayed the first stage of Stalinist collectivization. Because of immediate censorship, only a few copies of this original publication survived – the poem was perceived as a critique of Soviet agricultural reforms (Kahn et al. 2018: 609). It has been interpreted either as a satire or an utopian depiction of technological progress (Masing-Delic 1992: 272–3, n. 36; Cheloukhina 2013: 201). In particular, the distinct voices of farm animals in the poem complicate its meaning or at least preclude us from understanding this text as a mere commentary on the economic situation at the time. Rather, the animal characters exemplify the prophetic dimensions of this poem, believing that the transformation of the village is a stage in some kind of spiritual emancipation of non-human beings: 'an earnest prophecy – or at least a mental test of the idea – that collectivization and Communist science will free animals from exploitation and disease' (Ostashevsky 2005: 31). However, because animals speak in the poem, Zabolotsky was accused of mocking collectivization (Cheloukhina 2013: 203).

Agriculture Triumphant was probably not meant to be critical of Soviet politics, although Zabolotsky's repeated and prolonged criticism in other texts resulted in the poet being arrested and sent to concentration camps in Asia and the Far East (Brown 1978: 49). The poem was misunderstood since it translated the communist revolution into a language of salvation for humans and animals. Zabolotsky used tropes from communist propaganda about how 'feudal lords' and 'kulaks' mismanage and exploit animals (Masing-Delic 1992: 279) to portray the dawn of the old world, when animals' deaths were unimportant. Due to the major role played by animals in its vision of this new world, the poem invites us to reread it from an ecocritical perspective and understand it as both an eco-utopian and reformatory representation of the early stage of collectivization. To some extent, the poem expresses Zabolotsky's utopian beliefs and offers a poetic alternative to the ecological trauma that Stalinist reforms caused.

The title of the poem, *Agriculture Triumphant*, alludes to the socialistic newspeak that praised the development of agriculture not only as a branch of industry but also as the 'new Soviet man's' scientific transformation of nature. The Prelude opens with an image of a small village that does not look like it belongs to the world of rational and controlled nature, since 'a tree reels, nearly falling, / while a wispy river's scrawled there. / A few peasant huts are standing / by the loony little brook, / and an oblong bear is ambling, / late of evening, to his nook' (Zabolotsky 1999: 61). Nature's irregularities and disparities isolate this place from the Soviet realm, as well as from the new, rational, anti-natural farming technologies. These pastoral depictions of the village's landscape and environment suggest from the beginning that Zabolotsky does not follow the fixed formulas of Socialist Realism, which rejected any spiritual transcendentalism. Instead, he brings together two contrasting worlds – the old world of nature's mystic transformation and cosmic harmony, and the new world that the transformative promise of the old village's collectivization brings.

What makes this poem exceptional in a time of environmental crisis caused by collectivization is that Zabolotsky cannot imagine the transformation of the village without the liberation of the animals who, according to the poet, speak the language of nature. Therefore, some commentators emphasize that the poet links ideas of the 'interdependence of man and nature' (Goldstein 1993: 114) with a fascination for technological progress that is conditional on humans always taking responsibility for just relations with the natural world.

The revelation of what is happening in the village is announced by a bird. The crane heralds a new time to come with a declaration that 'fills the silent sky / with his whooping battle cry. . . / His beak unfurls a slogan: "Listen!" / cries the scroll: "The Three Field System's / Unproductive"' (Zabolotsky 1999: 61). In the first part, the poem freely recreates a discussion between peasants about the soul, which encompasses nature's 'fearful mess' (Zabolotsky 1999: 62). These peasants embody the old world and Russian Orthodox mysticism, but their role in advancing agriculture is the opposite of that represented by the figure of a soldier. Zabolotsky didn't choose a worker but a military professional, as if to stress the violent nature of collectivization (Cheloukhina 2013: 196) and the danger this new scientific approach might pose to humans and animals. Therefore, in the poem's second part, animals join the discussion and 'nature's soul [comes] within their [the peasants'] grasp' (Zabolotsky 1999: 66). In canonical Christian texts, animals are devoid of souls, but in the poem, they speak on their own behalf, caught between their poor condition as farm animals in the old world and the new modernized, mechanized village where they will not be needed anymore. These animals, however, do not resemble those in George Orwell's *Animal Farm*, though even the animals there do not function only as allegorical figures for the Russian Revolution. In fact, in the second part of the poem, under the pointed subheading 'The Sufferings of Animals', Zabolotsky's animals do oppose the men who represent the old world and their hierarchal relations based on power and violence. In parallel with Socialist Realism and its negative image of the *kulak* or prosperous peasant, for the animals the *kulak* represents peasants in general and the egocentric world of a 'blind feudal lord'. Here, for the first time in the poem, the utopia of socialist revolution is directly connected with animals' emancipation, and the *kulak* is the only obstacle to fulfilling revolutionary goals: 'The cottage stands there, like a fort, / snug inside the blighted world, / showing its owner how absurd / are kolhoz, freedom and hard work' (Zabolotsky 1999: 69–70). The question about how exactly animals might benefit from collectivization remains open, however. Evidently, it is an occasion for Zabolotsky to comment on unjust treatment of animals by humans. The horse in *Agriculture Triumphant*, for example, speaks in a didactic way: 'You are wrong, men, if you think / me incapable of thought, / when you beat me with a stick' (Zabolotsky 1999: 67). On another occasion, the horse is a creature endowed with some metaphysical knowledge: 'All round me nature's dying. / In its wretchedness, the world / falters, flowers weep, expiring' (67). This image of the created world passing away would stay with Zabolotsky, because he is said to be inspired by the Russian Cosmists' ideas, which influenced the majority of his 'nature' writings (Grossman 2000; Young 2012: 215), including this poem (Masing-Delic 1992: 273; Goldstein 1993: 125–31). Consequently, its narrative of transforming villages into a mechanized and modern chain of 'kolhoz-cities' (Zabolotsky 1999: 84) is strangely set in the wider, cosmic context of transforming the existence of all living creatures. This cosmic dimension complicates interpretations of the poem as either a satiric or utopian vision of collectivization.

If we assume that *Agriculture Triumphant* retells the achievements of collectivization as if animals were to benefit from it along with humans, the poem's vision of collectivization, heralded by the soldier, looks rather like a bitter satire. The fifth part of the poem seems to be the most important for transformation both on earth and on the cosmic level where immortal souls dwell. It is titled 'The birth of science' and explains all the 'good' the Revolution brings for non-human comrades. A 'Soldier', who is an ambiguous guide and narrator here, tells a story about how science rises 'above the world of bitter sorrow' where 'the bovine tribute swells' (76). The human and animal revolutions appear to merge: 'people and their cattle rise, / horses, oxen rise up too', linked by 'the red atom of rebirth' (76) spread by the Red Army. The atom is a metaphor for the tiniest particle that can be transported and replanted. It comes from Zabolotsky's fascination with science, especially with Konstantin Tsiolkovsky's philosophy and astrophysics (Goldstein 1993: 135–8), and from Cosmists' ideas of material continuity, such as Nikolai Fyodorov's concept of matter as 'the dust of ancestors' (Goldstein 1993: 131). The Soldier, a messenger of the scientific revolution, describes the future before the animals. This future, which will soon arrive and will destroy the old world of human and animal slavery, radically changes the status of animals from natural beings, with all the disadvantages of living creatures, into mechanized objects:

Cattle, listen to my dream.
In my sheepskin, I was sleeping,

and the horizon parted to reveal
a lofty Institute of Creatures.
There, the air was fresh and sweet,
and in the centre of that place
stood a fine cow, full of grace,
its consciousness still incomplete.
Goddess of Milk, Goddess of Cheese,
brushing the ceiling with her head,
bashfully she raised her frieze
and thrust her dugs into a keg.
Ten streams descended with a crash,
pounding the hammered-metal vat,
and the can, ready for dispatch,
blared like a military band.
And that ecstatic cow stood tight,
her arms folded across her chest,
game for anything that might
ameliorate her consciousness.

Zabolotsky 1999: 77

This super-efficient milk-cow is the embodiment of a soulless machine that represents the accelerated progress of agriculture, which is realized in industrialized animal farm production. The 'Institute of Creatures', like the real scientific institutes located in various Soviet collective farms to test new biotechnologies and machines, also involves the other animals, who, according to the Soldier's story, are preparing new food with the help of chemistry. The Horse, however, not convinced by the soldier's fantasies and thinking realistically, interrupts: 'soldier, have you heard the wheezings / of your poor, tormented ox?' (79) The Soldier, in turn, introduces the tractor 'who's crawling round / the hill' (79) to take the Horse's place and allegedly make his life easier – but in fact to render all horses redundant:

Metal, two-storeys high it stands,
with cast-iron snout and spitting fire,
master of combat, hand-to-hand,
with nature . . . it's coming here.
Take heart, you cows of intellect,
take heart, you bullocks and you horses,
henceforth, sanitary quarters,
for all of you we shall erect.

> Overthrowing plough and harrow,
> we'll raze the old world to the ground.
>
> 80

The Tractor Driver ends this forecast of the future with the elevating words: 'We are constructing a new world, / a brand new sun and brand new grass' (81). He does not say that these animals, who according to Zabolotsky possess soul, will be rendered obsolete by dairy and meat production and extensive cattle breeding.

This visionary poem presents a controversially utopian version of the 'triumph' of agriculture over the old village, which will affect human–animal relationships. Paradoxically, the animals that have been given voice here are the most critical of collectivization, since their future seems dystopian rather than ideal. Similarly, the peasants on the collectivized farm 'eat well' and 'peruse newspapers' (84), which sounds like a satiric poetic fantasy that was never fulfilled. It is possible that Zabolotsky, at this stage of his writing activity – he was under 30 – believed that he could adapt the language of Soviet propaganda to his own philosophical ideas, since the communist 'new world born in toil and pain' (82) is just another threshold in what he understood as cosmic evolution. This cosmic trajectory exceeds mankind's existence on Earth, and overcomes death and imperfect natural laws (Masing-Delic 1992: 281–6) in this closing vision of the plough as 'the queen of earth' deliberately abandoned:

> on the hill above the river,
> cemetery worms devour,
> for the very first time ever,
> the wooden body of the plough.
> So, the queen of earth is dead,
> that peddlar so beloved of crones,
> and rising high above her head,
> the burdock of oblivion!
>
> Zabolotsky 1999: 84

All in all, the poem does not praise agriculture as it was promoted by Soviet propaganda, but presents it as potentially triumphant in overcoming the old rural world. Starting with the Prelude, it is unclear exactly why the revolution begins. However, even as tractors roll out and industrial, rationalizing farm production is implemented, this revolution does not involve deeper reflection on the relation between a cosmic, spiritual natural order and biological death. Zabolotsky incorporates metaphysical ideas by taking up the utopian project of collectivization, though not in its Soviet interpretation. Instead, the poet satirically comments on Soviet collectivization and discusses how he understands the emancipation of all the bodies and souls of human and non-human residents of the village on a cosmic level.

In reality, 'collectivization was a traumatic experience for Russian peasants' and not comparable to previous state reforms (Fitzpatrick 1994: 4). It was 'more than [the] mechanization of agriculture and the creation of a modern agronomy to feed burgeoning cities. It was violent, brutal, and murderous coercion, a revolution of totality, rapidity, and violence' (Josephson et al. 2013: 97). It was traumatic for humans and animals as well. The first phase of collectivization included the removal of the peasants' livestock (Fitzpatrick 1994: 4) by the state, which not only caused an increase in hunger but also the senseless death of animals. Later Stalin admitted that it was a mistake (Fitzpatrick 1994: 63), but 'many collectivized animals died of mistreatment and neglect' (Fitzpatrick 1994: 4) because they could not be killed en masse in the slaughterhouses (Fitzpatrick 1994: 53).

In rural culture, peasants' attachment to farm animals originates from the triad of relationships amongst peasants, soil and animals, in which animals are mute but subjective beings, or at least living things on whom peasants are dependent. The relationship between the peasants and nature has been inscribed in patterns of interdependence and exchange in a kind of culturally-bound ecosystem, which means that animals are not only instrumentalized but are also often treated as companions, friends and members of the family. Comparing the rural world and its internal relations to an ecosystem is an epistemological metaphor, a metonym of hybrid community, that does not separate culture from nature. The pastoral image of traditional peasant culture is even based on a kind of 'exclusivity' because of its relation with nature, as captured in a poem *When the Village Daylight Dimmed* by Oleg Chukhontsev. The poet's collections were in majority banned from being published because they did not follow the rules of the Soviet propaganda. In this poem, particularly, the rural world exists outside the communist reality and the poet longs to 'share this world' though it escapes his means of expression:

> When the village daylight dimmed
> And the cock crowed half-asleep
> And the apple branches rustled,
> Then I thrilled to sense the presence
> Of a strange, exclusive world
> Which nature had revealed . . . .

> I longed to find the simple words,
> The artless idiom of nature;
> But the damp grass rustled
> And the black leaves murmured,
> Excluding me, and hinting faintly
> At another world, another language.
>
> Chukhontsev 1993 [1968]: 952–3

One of the reasons for this 'exclusivity' is that peasants are described in cultural memory as the ones who identified themselves with the cosmic rather than the anthropocentric perspective. They perceive themselves as one of many elements of God's creation; as a traditional and pious community, they also accepted their own limited and weak agency in relation to nature's laws (I will return to this theme when analysing *Konopielka*). This mysticized and romanticized image of the village inspired the greatest Russian poets, including Joseph Brodsky, who in the ahistorical village finds a special kind of non-Orthodox spirituality:

> In villages God does not live only
> in icon corners, as the scoffers claim,
> but plainly, everywhere. He sanctifies
> each roof and pan, divides each double door ....
>
> The chance to know and witness all of this,
> amidst the whistling of the autumn mist,
> is, I would say, the only touch of bliss
> that's open to a village atheist.
>
> Brodsky 1973 [1964]: 81

Within a specifically constructed peasant worldview, everything was created before humans, and human power is subordinate to creation. In this sense, the pre-revolutionary rural world, before collectivization, also represents a premodern, pastoral and self-sufficient world connected with the soil (Brown 1978: 220). A peasant village stands for an ahistorical equilibrium and a homogenic community isolated from urban areas (Parthé 1992: 6), as in indigenous societies where direct, material interdependence among humans and non-humans aims at survival. It is also a polyphonic world whose capacity for articulation was suppressed. The voices of this world are recreated in literature because it can represent nature's presence, the gaze of farm animals and the response of the soil. With collectivization, however, the peasants' 'pastoral time' is 'invaded by history' (Parthé 1992: 13) and this multivocal relationship with nature is destroyed, as in

Zabolotsky's poem *I Do not Look for Harmony in Nature* (1947) where 'the soul senses neither voice nor harmony' (Zabolotsky 1999: 177). The environmental costs of industrialization affect the poet, who sees the exhaustion of the natural world but also believes in industrial progress and nature's transformation: 'Around me nature's sad and heavy breathing. / And wild freedom and good mixed with evil / are not in nature at this moment. / It is a dream of glittering turbines, / measured voices of labour and reason, / the chanting pipes the pink glow of the dam, / electric power, human construction' (177).

Along with the ecocritical futurism of *Agriculture Triumphant*, Zabolotsky wrote ecocentric poems, wholly devoted to plants and animals, that belong to his 'stream of philosophical lyrics involving nature' (Brown 1978: 48). He seems to have experienced the Soviet industrial revolution in a paradoxical way, aware of its destructive impact on the environment as a sort of *nature morte*, but nevertheless seeing more life around him. In a poem written in 1936, *Yesterday, thinking about death*, he even considers himself to be 'Nature's thought! Her trembling mind!' (Yuzefpolskaya and Rueckert 2006: 291).[2] In his ecstatic, mystic poetry, he searches for perceptions of immortality in the natural world and the human mind (Yuzefpolskaya and Rueckert 2006: 293), and develops the idea that 'man could come to life again through nature' (Goldstein 1993: 144). But even in his realistic diagnoses of the environment's vulnerability, as in the poem *I Do not Look for Harmony in Nature*, nature never fully dies in Zabolotsky's poetic world (Goldstein 1993: 138). Or perhaps for him, nature 'was too alive' to die, as in Arseny Tarkovsky's memorial poem dedicated to Zabolotsky: 'For everything was too alive – the leaves, the / grass – / As if someone had placed a magnifying glass, / Over this world of embarrassed thrust, / Over this net of pulsing veins' (Yuzefpolskaya and Rueckert 2006: 291).

Animals have a special place in this world of 'pulsing veins'. Zabolotsky wrote some poems devoted to them out of his admiration for nature, for his native rural world, while others focus only on philosophical ideas. In the poem *A Walk* from 1929 – written in clear language unpolluted with propaganda, in the same period as *Agriculture Triumphant* – a sympathetic curiosity seems to prevail. This poem recounts a self-effacing but attentive observation of a bull, where an aesthetic view intertwines with philosophical and visionary thoughts:

> The animals have no names
> who said they should be given any?
> Suffering without end
> is their hidden destiny.

> The bull moves off into the pastures,
> talking with nature as he goes,
> over lovely eyes are planted
> alabaster horns.
>
> ...
>
> Why is she so melancholic?
> What is it that ails,
> the whole of nature smiling,
> like a cheerless cell?
>
> ...
>
> Before it, waters sparkle,
> and the vast woodlands sway,
> nature, with its laughter,
> dies each moment of the day.
>
> Zabolotsky 1999: 125

The poem veers between reflecting on the animal's condition and on that of the dying world of nature in general. It is a reading of a landscape – a narrative landscape poem. It begins with the question of naming animals, which is understood as possessing them through this naming, though that does not end their suffering. Outside the human world animals, even very common farm animals such as the bull, do not need names. The bull is depicted as beautiful, with 'lovely eyes' and 'alabaster horns', despite cultural representations of cattle bred for meat as terribly deformed by industrial farming. His subjective presence in the poem is nameless. The poetic filter creates a simple picture: a bull lying in the field, nothing interesting – a pastoral scene where time slows down and 'a bird circles'. But Zabolotsky, without mentioning humans in this landscape, detects 'sorrows', melancholy and something that 'ails'; then he sees the bull's tears and deciphers the paradox. This sad, lingering pastoral image, juxtaposed with a pre-agrarian era where 'the vast woodlands sway', displays nature as already constituted in the horizon of finitude. What, then, does this 'laughter' mean? The laughter of 'the stream', of nature, of history, 'a cheerless cell'?

The poem raises philosophical questions for which Zabolotsky was finding answers in Fyodorov's and other Russian Cosmists' ideas. The last words of the poem – 'nature, with its laughter, / dies each moment of the day' – apparently contrast with the conventional understanding of nature as a cycle of renewal. These final words oppose nature's signification of vitalism and progress, and specify the meaning of time: time's passing is perceptible as an irreversible but recognizable process happening every day. The laughter may relate to humans' naivety in denying that their world is finite or to their ignorance that nature does not ultimately die but recreates itself through death. The bull, a member of a species (cattle) condemned to be slaughtered, here looks like a tricky, distorted portrayal of a muse – a muse that indicates death in an ahistorical village. In the period of collectivization that Zabolotsky witnessed, however, farms were 'improving' their methods for rapidly rearing and killing animals. The poem, then, offers a rare representation of an individual animal decontextualized from the realities of industrial farming, but one who still 'thinks' he is condemned to die, together with the whole natural world. The philosophical idea of nature not as a dying world but a world of transformation not only eradicates the sadness of the poem, but also revives the tradition of nature representing life, even as the poet witnessed the fragility and vulnerability of nature under assault by collectivization.

In Zabolotsky's poems, farm animals are aesthetic and metaphysical figures. At the junction where environmental history meets cultural memory, farm animals inspire counter-narratives to the industrial production of meat and milk, resonating with the image of the pre-industrialized village where herds of cows safely grazed in the meadows. This ideal stored in cultural memory, which can still be found in Eastern European literature (e.g. novels by Wiesław Myśliwski), was first distorted by the collectivization of farms that took animals from their private owners, from the peasants who knew their needs, and subjected them to an industrial process of production or mass destruction (Josephson et al. 2013: 97).

In correspondence with Zabolotsky's *Walk* – albeit much later in 1980 during the late stage of communism in Eastern Europe – the Polish poet Julia Hartwig created a cycle of poetic portrayals of a cow that also oppose the image of industrialized animals being bred for milk and meat. In her poems *Portrait I, II, III*, Hartwig (1980: 12–14) addresses the cow as a friend ('my friend' or 'comrade'), whom she calls a 'native sphinx', a common but mysterious animal. This cow must be an exceptional being since she is inscribed into a portrait-narrative. Her distinguishing features are discovered not just by looking at the animal in the field, but by observing her in a way that avoids associating the cow with the industrialized image of an overexploited creature headed for eventual slaughter. This approach to observation, as in traditional peasant communities, anthropomorphizes the animal and adds a poetic element to this image. It questions negative stereotypes of the cow, which is presented in Western culture as stupid, boring and dirty: 'she hasn't believed that digestion is her only activity' (*Portrait I*). The poem takes a closer look, as Zabolotsky's poem zooms in on the bull – this cow is really pretty. She is depicted as an aesthetic object: she has 'a tongue similar to red orchid', and she resembles a 'Creole', 'thick' and 'shapely' girl (*Portrait I*). She dominates the landscape (*Portrait II*), but also looks at the person speaking in the poem ('seen eye to eye' – *Portrait III*). Historically, in relation to humans, a cow is simply a container of milk, but in the poem, she contains 'patience' and 'mystery'; she is a 'native sphinx' who cannot be wholly known (*Portrait III*). Because of the animal's tranquil nature, her slow and delicate movements, the pace of the poems is also slow: it has line breaks but the syntax is like that of prose. As a ruminant, she chews on 'her own destiny' with 'no hurry' in 'a perfectly close system of her interior'; the 'flowers' and 'grass' that grow again and again 'create for her an illusion of immortality' (*Portrait III*). While Zabolotsky's animals are unconscious of their immortality, Hartwig's cow participates in a cosmic peasant world, beyond humans' knowledge, in her own microcosmic and metaphysically renewing reality.

Hartwig's later poem *It will Speak* (2009: 36) is also framed as a portrait of a cow, but it is a pompous and funereal text about the voice of the animal (the titular 'it'). The cow moos – she speaks not in words but in emotional registers that can be easily recognized, read, understood and intercepted, and that include the deepest fears of passing and death. In this poem, the animal plays a double role. On the one hand, she is a philosophical figure for animals in general (as in Rilke's modernist *Elegies* or Agamben's *The Open*), who embodies the other and offers overlooked access to the key issues of our earthly dwelling, expressed in the poem through the metaphor of a 'useless musical instrument in the human choir' (her voice is that of a tragic, lonely hero in 'the world's meadow'). On the other hand, she serves as an exemplum of what humans and animals share, what is inscribed in the paradox of finite existence, and what is equally well expressed by her 'agonized' roar.

These poetic images of farm animals, such as the bull in Zabolotsky's *Walk*, the cow in Hartwig's *Portraits*, or the horse in Zabolotsky's *The Face of the Horse* (1993 [1926]), originate from the trope of the isolated world of the old village. Here, 'isolated' does not mean forgotten, but rather, through its strong bond between humans and animals, the village resists Soviet colonization of nature and gives voice to a more-than-human community. The metaphysical content of the poems, whether Zabolotsky's Cosmist ideas or Hartwig's account of the cow's otherness, creates a distinctive niche in cultural and environmental memory, a counter-narrative to the village's transformation during the long period of collectivization. Through these counter-narratives, collectivization does not only mean a change in managing agricultural production in state-led Soviet economies. It can be perceived as a cultural phenomenon in which an instrumentalist worldview encroaches on human–animal–environment relations and reorganizes the triadic rural world.

The dying old village, affected by collectivization and communist agrarian reforms, had already emerged as a prominent theme in Soviet literature in the 1950s, and it continued to feature in novels and short stories published in the two succeeding decades. Especially in the Thaw period, in the 1960s and 1970s, writers who declared themselves against collectivization (Hosking 1989: 57; Parthé 1992: 67) are associated with Village Prose (Clark 1989: 76–7). Among the better-known examples of this genre in Russian are Aleksandr Solzhenitsyn's 1963 story *Matryona's Home* (Parthé 1992: 19–20, 45–6), and works by Fedor Abramov, Valentin Rasputin, Aleksei Leonov and Andrei Belov. They primarily emphasized the loss of spiritual and ethical values associated with the passing of old villagers, but because they are critical of collectivization, their texts can be read as ecologically-oriented prose as well.

A representative Village Prose text from the Thaw period is Boris Mozhaev's *Lively* (2008 [1966]) – but it is still contaminated with ideological language, which enabled its publication. In this story, Mozhaev tells of a rebellious peasant who quits the *kolhoz* and wants to survive outside it, despite the community trying to force him to come back. With his meaningful name, Lively serves as a metaphor for peasant life organized by the human–animal–soil triad, as against death in collective farms. He represents a shift in Russian village literature towards severe criticism of the introduction of *kolhozes*, understood as 'the brutal destruction of the organic order of peasants' lives that causes justifiable indignation, resistance, and eventually rebellion' (Kahn et al. 2018: 736).

Devastation of the peasant village as a model of culture is a topic in Russian prose as well as in the poetry of Vladimir Soloukhin, who himself was born into a peasant family and who 'complains' in his works that he has to hide themes of love in his work with representations of collectivizing agriculture (Parthé 1992: 14). His resistance to the idea of progress, typical for the Village Prose writers (Parthé 1992: 26), was more forcefully and radically expressed in his poems, such as *To Make Birds Sing*. It is a love poem but written with an awareness that reality has completely changed – in metaphysical, social and environmental senses – following the communist Revolution, when 'the birds are not able to sing'. Using the phrase 'his post-decembral kingdom' to refer to Stalin's regime, the poet faces the reality of terror and disappearances of people: 'I cannot calm the blizzard, / Or melt away the snows. / But it is in my power / To make birds sing' (Soloukhin 1993: 753). He therefore returns to the pastoral context of rural poetry, but at the same time he cannot do so without redefining the pastoral (Gifford 1999). In this case, nature can no longer be idealized because of the transformations of the Soviet era, but it can be preserved in poetry, as if poetry has become a space for the conservation of cultural-environmental memory thanks to love.

Redefining the pastoral can be considered 'a hallmark of village prose' as well. Katerina Clark writes that 'its major theme [is] of how the "machine" as, variously, urbanization, Sovietization, or the sense of alienation and the loss of the old values and standards which comes with the erosion of the *Gemeinschaft* world, is destroying the rural "garden"' (1989: 83). The rural world represents a microcosm not only of Soviet society, but also of its problems during modernization, industrialization and collectivization. Reviving peasants' voices, perceived as backward in dominant Soviet propaganda discourse, and reading them ecocritically, gives an argument that peasants' attachment to traditional morality of the village and their reluctance to 'progress in technology' (Clark 1989: 91) can be understood as environmental resistance instead.

An illustrative example of this in Polish Village Prose is the novel *Konopielka*, written by Edward Redliński in 1973.[3] Although the novel was not censored by the communist authorities in Poland – it was even filmed in 1981 by Witold Leszczyński – it is read here as an ecocritical example of a peasant community's oppositional response to human intervention in the river's landscape. However, when the text was released, it was understood at the time as critiquing the peasant village's ignorant obstructionism, full of superstitions and stupidity, not the communist effort to drain the wetlands.

The rural community lives there as if in an ahistorical space on eastern Poland's outskirts, driven by respect for a divine order where God is not provident but omnipresent: 'the forest hears, the field sees and there is no place where you can hide because everything looks at you' (Redliński 1973: 75).[4] The peasants' beliefs are radically constant, even when they are faced with tragic events. When the main character's daughter dies, no grief is expressed and there is no time for mourning: 'Jadzia is dead. The winters are cold, the summer hot. There are fish in the river. Birches in the forest' (113). People accept their fate rather than opposing it. Their isolation is also strengthened by their rural dialect, which is only oral, and by the remoteness of their area.

The village's name, Taplary, echoes the conditions of a wetland environment – the verb '*taplać się*' means to get wet and dirty when playing in the mud. The residents of this village have an affirmative attitude towards their muddy, sodden abode as well as towards the whole world created by God. The wetlands are located on the River Narew, where people are accustomed not only to the surrounding mud but also to the river's floods. When the local communist authorities propose modernizing the river bed and the wetland area, the peasants do not understand that the swamp can be drained (42). Therefore, these plans for modernization of riverside areas meet with their strong opposition. They also believe that order in the world is taken care of by '*Pambóg*' (the Lord, in their dialect), but when He has to rest, 'then the devil does what he wants: one finger to roll, to change, to change!' (40–1) – because only the devil wants to 'change' and 'improve' things (40). In their opinion, they are defending the village against evil, against secular officials who 'do not like that there is a village, which still lives a traditional, godly life!' (42). Their rural traditions and the wetlands are inextricably linked, so destroying one destroys the other.

It does not help when the authorities have a teacher, who is new to the village and who is compared to a '*konopielka*' (the religious Easter song proclaiming good news), to explain the necessity of ameliorating the river bed and drainage of the swampy area to the peasants of Taplary. For them, a flood is not a disaster but an ordinary, unchangeable part of their local identity. Triumphing over the elements, as it was presented in classic Socialist Realism (Clark 1989: 94), is here reversed and depicted as the unintentionally pro-environmental consciousness of a 'backward' peasant community. Therefore, they do not react when the mayor screams at them and calls them backward savages: 'Do you really want to live your whole life in those rushes like ducks, you and your children? . . . in a year the mountain under the Bokinami will be dug out: the water will flow and there will be no trace of the old river beds, distributaries, wetlands, bumps; there will be only one main river left under Suraż' (Redliński 1973: 55). The locals, though, do not care about the vision of a modernized area and life. In subsequent meetings with the authorities, dehumidification and melioration are explained to them, and plans for a road and the construction of electricity pylons are presented to them as well (Redliński 1973: 147), but they are intransigent: 'we don't need melioration and electricity, we feel good in the mud and got used to it' (148). Paradoxically, their persistence resonates with today's initiatives to restore and protect the wetlands as natural guards against distant flooding. By protecting the Narew ecosystem's old river bed, they are unwittingly helping save the whole river from intervention – and it largely remains one of the most meandering, unregulated backwater rivers in Eastern Europe, despite destructive, techno-oriented hydroengineering and industrial pollution that affected other rivers during the communist era.[5] Eventually, when authorities implement some of the modernization plans, especially by draining the river's tributaries in some sections, the peasants in *Konopielka* are shocked. Drainage not only demonstrates how rich the river was in fish, as children collect baskets of pike, orfe, perch and roach fish from the drying river bed, but most of all they find it unbelievable that such a huge river, 'their river', could disappear: 'people are crying ... fish are lying in the mud as in oil, dying in the sun. Oh, *Kyrie Eleison!* There was a river, and now there isn't!' (104). The disaster is painful when it is caused by people and not by nature.

The endangered rural and environmental world of *Konopielka*, when analysed as a cultural phenomenon in the context of contaminated language and idealized human–animal–environment relations, belongs to a vast imaginary of Eastern Europe as a peasant landscape. This image is romanticized in cultural and environmental memory by writers who are biographically connected with the village but who, especially in post-communist texts, reject a pastoral image of it, as the old rural culture is obscured by the Soviet past:

> The barn is gone and the orchard has been felled. It used to shield the house. Now the winds hurtle in from three corners of the earth and batter at the wooden walls. Especially from the east, from over the river. They blow in from the plains on the far bank, glide across the waters and stream onto the plateau on the other side. The eastern gable wall is no longer strong enough to stop them. They surge inside, into the two rooms we used only when guests came. The orchard is gone, along with the apple trees and their voluminous branches which bore the blast of the winds and protected us against heat and cold. ... a rural world in which houses were built of timber darkened by the weather, roofs were made of hay, and animals lived side by side with human beings. ... In the mid-1970s, the Druzhba ['friendship'] Pipeline ruptured. Black gunk came floating down the river. It settled on the banks, and on the reeds and willows growing beside it. One day someone set fire to it all. The thickets burned. The river was in flames. It was like war – but without the soldiers. I knew that the oil had come from Russia, from the Soviet Union. I knew this was in the East, far but also very near. I imagined Moscow as something distant, grey, huge and dull. It was unreal, yet we lived in its shadow.
>
> Stasiuk 2016

This translated fragment, taken from Andrzej Stasiuk's essayistic book *East* (2014), written in Polish during his travels across post-communist Europe and Asia, evokes the landscape closest to the author, the landscape of his childhood. It belongs to an old rural world that was wiped out by Soviet Russia.

Because of the Soviet collectivization, the myth of the rural village was disturbed in Eastern European cultural memory. It is no longer a representation of melancholic peasants who are resting in the evening accompanied by their animals or watching the storks while lying in the grass, as in Józef Chełmoński's (1849–1914) paintings (see Figures 3 and 4). This 'peaceful' and ahistorical village is gone; however, what resists in the memory of rural culture still refers to some pastoral tropes. Literature from the Soviet period involves representing the human and animal fate as shared due to collectivization, since collectivization threatened the whole human–non-human communities.

Figure 3 Józef Chełmoński, *Indian Summer*, 1875. National Museum in Warsaw. Public domain.

Figure 4 Józef Chełmoński, *Storks*, 1900. National Museum in Warsaw. Public domain.

Radical industrialization of the village during communism exposed how undermined the relations between peasants and farm animals had become. One of the responses of this period literature was to poeticize the relation with nature and the animal world, and escape from the brutality of what collectivization brought to rural traditional culture. Redefining the pastoral myth put the image of peasants' community in a new light: in their ability to resist communist governments' intervention by defending the non-human, metaphysical order.

4

Satantango: Interconnecting Human and Ecological Worlds

Though collectivization in Hungary was less intense than in Stalin's Russia, it still took place. Huge industrial farming conglomerates were set up and competed with small private landholders (Fenyo 2000: 313). They had their period of prosperity, but in the end they did not survive the political crisis and transformation that brought about the collapse of the Soviet regime. The late communist village provides the setting for the novel *Satantango* written by László Krasznahorkai in 1985, which I find consistent with Béla Tarr's 1994 film adaptation (the screenplay was written by them both). The decay of a collective farm's community offers a narrative that contrasts with that of propagandist newspeak, but not in the activist sense of Village Prose writers – this counter-narrative is full of despair. The text of the novel, written without paragraphs, and the film's lengthy scenes encapsulate what I call 'a tired village' – tired after decades of collectivization and an unsuccessful shift to industrial farming. Devoid of morality and portrayed in a wrecked surrounding, the community's vulnerable residents betray the very deep crisis and decline of the Soviet Union.

The social breakdown of *Satantango*'s village is depicted in this small rural community where nobody works. People drink in the local pub, cheat on each other; children don't go to school and adolescent girls work as whores. The mud, the neglected and shabby surroundings, and the immorality of the villagers do not necessarily come from poverty and despair. The residents represent common social problems for the late collectivized village before the collapse of communism: unemployment and lack of initiative. They are waiting for their messiah (Irimiás), who eventually turns out to be the communist secret service spy. However, the Soviet economy has affected not only people but also the landscape, and this is what *Satantango* communicates: from the rain, wind and mud that seem to be almost characters, through the village doctor's geological deep-time fantasies, to the graphic scene of animal abuse and countervailing scenes of insects, birds and horses as magical or liberated figures. Environmental history intersects with cultural memory in this late Soviet village filled with immorality and depravation, but at the same time human decay and immobility is placed in a multivocal space, as if beyond morality, in the vastness of nature.

The Great Hungarian Plain, which accounts for more than half the area of the country's flat, rural territory, is called 'the agricultural heartland of Hungary' (Danta 2000: 261) and is one of the most representative cultural, or even national, landscapes of Hungary (Batori 2018: 13, 31–2). Because of its ecosystem, it is described as a grass steppe and treeless semi-desert (Batori 2018: 31) that is often termed *puszta* in Hungarian, which means 'wild, deserted, plain land'. As such it represents spaciousness, limitlessness and freedom (Tuan 2014: 52). According to Anna Batori, the author of *Space in Romanian and Hungarian Cinema*, to set any action there means to refer to Hungarian history, to 'the national yet colonised space' (2018: 145) of the Soviet era (see Figure 5).

In this big flat grassland, the spacious and quintessentially Hungarian national landscape, Krasznahorkai located *Satantango*'s inactive, unkempt but still inhabited late Soviet collective farm. At the end of every October, the farm and its surroundings become a rainscape covered by 'the stinking yellow sea of mud' (Krasznahorkai 2013: 3) and isolated from nearby towns, railways and 'the

Figure 5 Endless grasslands (*Puszta*) on the Great Hungarian Plain. Photograph: Beroesz. Wikimedia Commons.

outside world' (64). From the beginning of the novel, and of Tarr's film adaptation as well, the rain plays an important role. It trickles down and makes people wet (Krasznahorkai 2013: 12); it changes from a drizzle into 'a veritable deluge, like a river breaking over a dam, drowning the already choking fields' (12). It pours without pause, 'the constant rain' (Krasznahorkai 2013: 64), 'the product of some hidden intent' (114–15); it can make the village 'vanish under water like a ship that had sprung a leak sadly proclaiming the pointlessness of the miserable war between rain, earth, and man' (115). 'The stories we tell about rain are also the stories rain tells about us,' writes Lowell Duckert in the essay 'When It Rains' (2014: 115). The rain has its own agency, but the powerful and resonant wind is expressively active in the film as well. The same goes for mud, which covers everything and 'kills off all forms of life' (Krasznahorkai 2013: 64). Because of its vastness, the *puszta* is described as a monstrous 'muddy ocean', possessive towards the surrounding landscape and frightening as if 'breathing' (41). In *Satantango*, these elements of the weather speak and have something to say, as if they are playing the role of witnesses to the crisis and near fall of communism, jamming and obscuring the ugly reality of this collapsing estate, or even wiping this bleak human history from memory. It is then not a coincidence that Krasznahorkai and Tarr create a space that the novel and film share. These media complement each other to represent a vibrant world where landscape and weather are inseparable from the history of this space, from their 'environmental agentism' (Cohen, Duckert 2015: 6): the environmental elements intervene in this late-communist history because they are loudly voiced by the author and director.

The story begins at the end of October with the long autumn rains that persist, with a break for winter's snow, until late springtime. The rain fills the space and introduces many 'wet' metaphors into the language of the novel. Residents of the farm feel that they are rotting in this place (Krasznahorkai 2013: 9): their filthy life is like 'this filthy weather' (17), but they are immobilized. Their buildings need renovation and limewash (16, 19), 'the gardens [are] overrun by weeds', their industrial equipment has been stolen, and the whole area 'looked like a bombed site' (151), although they are still waiting for something. But there is a deeper crisis that is hidden from them – 'if they read the papers properly they would know that there is a real crisis out there' (Krasznahorkai 2013: 35) in the communist state. Then the charismatic and commanding Irimiás and his companion Petrina manipulate the villagers' apathy and take advantage of this situation of decline. They have been working as inside men for the local authorities and, from their perspective, peasants are just slaves who need a master, otherwise they 'go mad': the master can change but the peasants' mentality stays the same (43).

Near the end of communism, villagers who stayed in their inactive and deserted collectivized villages were a social problem because they felt not only useless and marginalized but also passive and helpless (Feshbach and Friendly 1992: 54). It was a challenge to make this group adaptable and active. To some extent, *Satantango* reflects this – in, for example, the broken alarm clocks (Krasznahorkai 2013: 13) that do not wake the peasants up to go to work because they do not work, except the prostitutes; in the vicious, immoral and cruel relations amongst the peasants (46–7); or even in the way that Irimiás easily influences the peasants' life decisions, mesmerizing them like a demon. Fiction and the crisis of the late communist village coalesce here, as it further damaged social relations and existence itself. The recollections of an anonymous peasant, quoted by the historians Feshbach and Friendly, resonate with *Satantango*'s atmosphere: 'We were all supposed to be just one big family after collectivisation. Just the opposite! Everyone was pitted against everyone else, everyone suspicious of everyone else. Now look at us, a big stinking ruin' (1992: 55). It is the same in the novel, where the residents of the estate do not trust each other but instead betray and spy on one another, 'peeking through their curtains to keep an eye on affairs outside' (Krasznahorkai 2013: 8). Outside they are surrounded by a vast muddy ruin, which only underscores how dirty and alienated this environment is.

Only the village's doctor, who watches others, too, is a different kind of observer. He spends most of his time behind his window curtain, sitting in his armchair, smoking and drinking Hungarian brandy, *pálinka*. In the film we can see how enormously fat he is; he does everything very slowly and with great effort. In one scene, which lasts 37 minutes, we are in his house, and he barely moves. The quiet long-takes are only partly filled with his narrating voice. He constantly takes notes about the peasants' activities, either about their routines or new events, in a systematic and ordered way – he has notebooks headed with each of their names. In the novel, though, we can understand his reasons for doing this better. His failing memory changes this seemingly senseless activity into a narrative that preserves the information he collects 'against the decay that consumed everything around him ... in the face of the power that ruined houses, walls, trees and fields' (54). Thus, he is a paradigmatic figure, as if he embodies the writer and film director hidden behind the stage and the drawn curtain.

The doctor's real-time observations are intertwined with fragments of geological history that the doctor reads. These fragments, which are italicized in the text, begin with the Permian period, when the first mass extinction of species occurred, and tectonic activity in the Mesozoic Era in Central Europe, including the genesis of the Great Hungarian Plain. These fragments represent *longue durée* history – which exceeds human imagination and precise data collection, and where human history is just a tiny blip in the Earth's natural history – but for the doctor, deep time is materially localized around him when he tries to imagine the sea that covered this part of Europe (Krasznahorkai 2013: 49). Dramatic geological transformations that also affected inland Hungary defined this area's landscape millions of years ago:

> The tensions within the earth's crust seek an equilibrium which does in fact duly follow when the unyielding inner mass that had hitherto been the determinant begins to collapse and sink, thereby bringing into existence one of the most beautiful basin groups in Europe, and as the sinking continues, the basin is filled by the neogene sea ... when the great lowland sea had mostly subsided leaving a large shallow lake roughly the size of the Balaton, how much destruction was caused by the combined forces of the wind and water.
>
> 55–7

The period when the European lowlands developed – when most of the modern taxonomic families of species, including mammals and hominids, already existed – is described as a natural catastrophe. The elements of air and water, strong winds and heavy rains still operate here, in this area of small lakes, bogs and marshes – 'the signs of the lost inland sea' (Krasznahorkai 2013: 58). Thus, the most distant past is described as if it were happening in the present. The doctor feels lost in not only geological but also prophetic history because of the captivating way that the book he is reading is narrated: 'the history of the earth that had seemed so solid, so fixed under and around him, came alive, through the unknown author's awkward, unpolished style – the book being written now in the present and now in the past tense' (58). This method of narrative, however, enables Krasznahorkai to create the bewildering realization that this actual place, the Great Hungarian Plain, had, a million years ago, been covered by the Pannonian Sea, which does not exist anymore. The large-scale tectonic and hydrologic changes were dramatic in character; although they belong to the geological past, they have left some visible traces that remain on the landscape and affect its weather patterns and climate.

This deep-time perspective leads the doctor to a special kind of alienation from the current situation of collapsing communism and corruption in the village. The geological past seems to be more real and engaging than human history, which is short and meaningless. The doctor cannot resist imagining 'himself as the defenceless, helpless victim of the earth's crust' (58) among other non-humans trying to escape from catastrophic land transformations: 'running, part of a desperate, terrified stampede comprising stags, bears, rabbits, deer, rats, insects and reptiles, dogs and men, just so many futile, meaningless lives in the common, incomprehensible devastation' (59). His imagination intensifies the illusion that even now, via a sort of trans-geological experience, he hears 'howls of pain … incapable of distinguishing between the general noise and ancient prehistoric screams that were somehow preserved in time ("The evidence of suffering does not disappear without trace" – he hopefully remarked) and now were being raised by the rain, like dust' (60). The real noises of 'whimpering', 'wailing' or 'stifled sobbing' in the present emerge and submerge, 'like the houses and trees that were solidifying into blotches', to 'rise clear of, or sink back into, the monotonous hum of the rain.' He calls this experience 'cosmic *wirtschaft*' (60), as the German term *wirtschaft* originally connoted 'hospitable reception and care' and to 'give and take the necessities of life' (Tribe 2015: 33) in a cosmic perspective beyond human knowledge and imagination. The landscape of sounds, the patchy surroundings within reach of the senses, the earth's history and time – they all swirl around in the capacious consciousness of the doctor, always a bit drunk but sober enough to halt the chaotic and unpredictable events for a moment in order to write and deposit them in his notebooks' memory.

Scaling up time and space puts the political and social situation of the villagers' world at a distance. Natural history intersects with human history in a universalizing experience of suffering that lacks an anthropocentric index. Even in the doctor's ordered activity of taking notes, a correlation between the narrative and the environment – the rain and dank smells – can be sensed, perhaps in some non-representational understanding: 'more-than-human, more-than-textual, multisensual worlds' (Waterton 2013: 66–75). The human and ecological worlds are strangely connected on the farm, which has been worn out by the long rain and decades of collectivization, but they are also inscribed in these deep-time events that are linked as dramatic repetitions in the doctor's imaginative memory.

Different aesthetic strategies are employed to render this bond, including a narrative structure that seems to mimic an endless tango, full of drama and passion. The book consists of two parts and each part of six chapters, like the six steps forward and back in classic tango's rectangle. At the end of each sequence, the dancers return to the starting point. The same applies to the numbering of chapters, which goes from chapter I through chapter VI in the first part, and then goes backward from chapter VI through chapter I in the second part, as if it were a dance in a circle. The title of chapter I of part 2, which is the last chapter of the novel, is 'The Circle Closes'. This final chapter closes but at the same time circles around. Fittingly, the first two and a half pages of the novel are the same as the last two and a half pages, because they are said to have been written by the doctor, which gives the impression 'that the novel itself is nothing else but a quote from the doctor's diary' (Kovacs 2013: 128). The film does not start with this text, but with a long prologue instead, possibly because 'it is very unlikely that viewers will remember after more than seven hours just what the exact text was somewhere at the beginning of the film' (Kovacs 2013: 128). Thus, this structural circularity is about returning to the same situation – when a change for the villagers becomes possible with the return of the satanic character, Irimiás – but then falling back into the repetitional misery captured in the tango metaphor. At the end, the peasants are still senseless, helpless and isolated by the constant rain and dirty mud spread everywhere, as at the beginning.

In Tarr's film adaptation of the novel, the landscape is 'one of the main characters', according to the director in an interview (Batori 2018: 148). He devotes lengthy scenes to portraying the Great Hungarian Plain in the rain, in motion, with a few bald trees and muddy roads, as the noise of the violent wind, or an accordion, interrupts people's conversations and drowns them out. The opening seven-minute-long take shows, in black and white, a vast muddy area behind some unkempt farm buildings, from which a herd of cows walk out. They loudly moo from time to time and slowly move towards the camera – one of them even approaches very close, staring in the camera's eye for a moment. Then they spread out among a village's abandoned-looking buildings, and the animals are left alone. From the beginning of this more than 7-hour-long film, the viewer may either be captivated by its 'slowness' or perceive it as too unbearably heavy and intense to watch – or both.

Indeed, Tarr's films exemplify 'long-take cinematography' (Koepnick 2017: 2) or 'slow cinema' (Jaffe 2014; De Luca seven Barradas 2016; Çağlayan 2018). The latter label does not directly refer to environmental criticism and environmental concepts of slowness, although it could be more put in relation with them, but emerges from a slow–fast polarity in the film industry and aesthetics that is rooted, for example, in Michelangelo Antonioni's films (Çağlayan 2018: 4–6, 9–10). Emre Çağlayan construes his aesthetics of slowness on the margin of dominant culture as highly valuable: 'in a world under rapid transformation and marked by an increasing pace of consumption, slowness is a marker for genuine taste, authenticity and wisdom, characteristics that situate slowness at the top of the hierarchy of cultural production' (2018: 8). He explains how he understands the extremely long takes and radically undramatic action in Tarr's films: 'while the limits of long take and dead time in slow cinema often extend beyond narrative function, I would argue that they retain, at the very least, a self-reflexive, conceptual relevance' (12). Further, 'the longer the scene pauses, the longer we engage in scanning the image for details that we may otherwise miss' (65). Lutz Koepnick makes similar arguments, naming 'Tarr's long-take aesthetic as a sign of intensified realism ... privileging the sensory over meaning' (2017: 102). The lack of action and the lengthy takes in *Satantango* definitely intensify the viewer's involvement as the camera moves in almost real time, although the film does become very dynamic when the weather changes. In one such scene, the camera follows Irimiás and Petrina as they walk through the town. Their monotonous walking is accompanied by a strong wind blowing a lot of garbage, especially papers, around them, which lasts for more than two minutes while nothing else happens.

The ugly, dirty and unpleasant setting – the peasants wear wellingtons all the time – is historically contextualized and visualized in the opening scene: 'what used to be an old farming collective is now being overtaken by domesticated animals in the total absence of human agency – visually as well as aurally' (Çağlayan 2018: 65). Batori, in turn, identifies the tendency of Hungary's post-transition cinema (1987–95) – the so-called Black Series that includes Tarr's work – to use black-and-white images and deserted environments to illustrate the multilayered crisis caused by the change to the economic system (2018: 15; see also Györffy 2001). For Batori, Tarr's *Satantango* is an emblematic example of the Eastern European universe looking like it has been 'absolutely destroyed, a territory no longer occupied by any foreign [Russian] power' and, in this film particularly, 'the rural dwellings and environment [seem] as wrecked as possible' (2018: 150). In the Hungarian *puszta*, where the Soviet communist economy left collectivized villages in such ruin, how strange and unreal the words of a leading Hungarian poet of peasant origins, Gyula Illyés, sound: 'People of Hungary, / stepchildren of history, / rise up out of the *puszta* / to reconquer your homeland!' (1971 [1932]: 17). The community of peasants in *Satantango*'s village is socially and morally destroyed, and cannot rise up again by itself: the landscape of the Great Plain can no longer be viewed as 'passive and mute' (Pick and Narraway 2013: 8).

The human and ecological worlds are entangled not only in this powerful representation of a landscape that often overshadows human problems, but also via the equally strong animal presence in the desolate world of Krasznahorkai's novel and Tarr's adaptation. The animals' role in the whole narrative is ambiguous to say the least: sometimes they are vulnerable, at other times resilient and liberatory. The text's most extended scene of animal abuse (Krasznahorkai 2013: 117–24) – which, in the film, is long and disturbed enough to be considered controversial and unethical – involves the victim–executioner relationship. But can a child be a murderer? Little Esti (Estike in the film), whose mother drinks, whose sisters prostitute themselves, and whose brother steals, is spurned by all of them. She does not go to school; she just wanders in her shabby clothes around the farm's shanties. Only the doctor notices her, soaked in the rain in a thin dress, neglected and miserable (Krasznahorkai 2013: 73). In the peasants' community, she is called retarded and weak-minded; orphaned by her beloved father, she is cursed by her mother. In the most cruel way, Esti tries to regain power over her life by torturing her cat, who at first trusts her and lets her do a lot of harm, only to be killed in the end by the rat poison that her coldblooded mistress gives her. During the crime, Esti rapidly transforms from a victim into a persecutor: 'the consciousness of her own inexhaustible grandeur ("I can do anything, absolutely anything with you...!") confused her a little at first, presenting her with a completely unknown universe, a universe with her at the centre' (120). She does not hesitate even for a moment, nor does she regret what she is doing at the time or afterward. In both the novel and the film, the cat's experience is not hidden – it is narrated and screened – and the torture is prolonged. Because rat poison acts slowly, the scene of the cat's death is also drawn out for long minutes: 'she had only to wait for the noise to die away' (124). Afterwards, Esti is incapable of thinking what to do next. Rejected by adults and her siblings, she leaves the village with the dead body of her companion – 'You're coming with me' (130) – and poisons herself as well.

According to Kovacs, Estike's death is given greater symbolic importance in the film's narration than in that of the novel because of the way the film reassembles the parts of the story differently. In the 'killing' scene, Tarr uses his 'most frequent technique to slow down the narrative ... [it] is the following of an action sequence in all of its most insignificant details. This creates a sense of *radical continuity*, meaning that virtually no element of an action sequence is omitted through the continuous representation of the given action sequence. *Satantango* abounds with scenes of this kind' (Kovacs 2013: 114). But only this one shows real animal abuse and demonstrates the horrific degeneration of the peasants' community, since no one there sets any rules.

There are, however, other creatures that play a role in depicting this run-down collectivized village near the end of communism. In the estate's inn, where unemployed residents get drunk, persistent spiders keep reweaving their 'fine' cobwebs after the last customer is gone (Krasznahorkai 2013: 148). The innkeeper cannot catch them because they cannot be seen; every night the situation repeats itself and he has to remove the webs. After a night of heavy drinking and dancing, when the peasants are waiting for Irimiás and Petrina, they fall asleep in the pub. Only one man is awake, playing the accordion, but the spiders come out as usual:

> The velvety sound of the accordion stimulated the spiders of the bar to a new frenzy of activity. Every glass, every bottle, every cup and every ashtray was quickly veiled over with a light tissue of webs. The table and chair legs were woven into a cocoon and then – with the aid of one or another secret narrow strand – they were all connected up, as if it were a matter of some importance that the spiders, flattened in their secret, remote corners, should be properly advised of every slight tremor, each microscopic shift, and would be so as long as this strange, all-but-invisible network remained intact. They wove over the faces, hands, and feet of the sleepers too, then, lightning-quick, retreated to their hidey-holes so that given one barely perceptible vibration, they would be ready to start again.
>
> 158

In contrast to the realistic scene of the cat's murder, the repetitive and persistent action of the spiders serves as a timeless metaphor for human latency. The spiders are presented as liminal creatures. Invisible to people, they exist between the known and unknown, but they must materialize their presence in the web before they can disappear. Thus, they interfere with the human world and the real world. By crossing the borders between human and non-human worlds and between reality and fiction, the spiders weave their all-encompassing net and structure the story of decay. Kovacs calls this 'the web-like structure' of the narrative (2013: 130).

The flies play a role similar to that of the spiders; the innkeeper, carrying a cloth, chases them for whole days. The flies soil everything, although they cannot be detected when they do it. Batori notices that the film's perspective and the movement of the camera resembles their flight when Tarr 'descends to the narrative space and joins the protagonists during their activities while examining every pore of their faces in long plan sequences' (Batori 2018: 155). He also lets 'real' flies sit on the camera's eye, as if to reveal even the smallest bit of dirt. The spiders and flies co-create the exhausted and destroyed world of the collectivized village, but they add a metaphoric layer to it; they make this history look not just bleaker but also more abstract and universal. From their non-human perspective, this ruined human world is encapsulated in rottenness and stripped of history, as if historical time had stopped while communism was disintegrating. In fact, even in the late 1980s, nobody believed that the Soviet Union would collapse. Pessimism prevailed, and people remember how they were devoid of hope even as the economic crisis reached a critical level. The novel and film represent this sense of ahistorical suspension, capturing the way no one felt that communism would fall even as it was breaking down.

However, not all animals co-create this wrecked world. In *Satantango* there are countervailing scenes of animals as magical or liberated figures, who either do not belong to the destroyed world of the collectivized village or are able to escape from it, which may anticipate the Hungarian *puszta*'s eventual decolonization after Soviet occupation not only in a social, economic and cultural sense, but in an environmental sense as well.

One example involves a long take of a white owl sitting on a perch in the old ruined manor at night. What begins as a white dot becomes a close-up of an amazingly rotating head as the camera zooms in on the owl for two minutes. The majestic creature emerges out of the darkness and enlightens the whole scene, as if stopping everything around it. In another liberating scene, which also metaphorically breaks the circular structure of the novel by presenting an independent story, horses escape from a slaughterhouse and disperse into a deserted town square. The town barman relays the news that a whole stable of horses are running free nearby and have not yet been caught in a short conversation with Irimiás (Krasznahorkai 2013: 221). Not long after, Irimiás and Petrina come across them: 'there they were in the middle of Eminescu Street, some eight or ten horses, grazing. Their backs reflected the faint streetlights and they carried on peacefully chomping the grass until they noticed the group staring at them, then suddenly, it seemed in unison, they raised their heads, one neighed, and within a minute they had disappeared down the far end of the street' (223). As portrayed in the film, the horses appear unreal, magical and full of a kind of beauty that is missing in the gloomy, ugly, dirty reality of the village. The horses have managed to free themselves, unlike the peasants, trapped and uprooted after collectivization, who are easily manipulated by Irimiás. In the film, they swiftly run into the square, their hoofbeats loud and powerful: they are the dominant creatures. In both the novel and the film, the animals' aesthetic aura prevents this damaged reality from falling apart. Even in the darkest times, 'Animals do not sleep. At night / They stand over the world like a stone wall' (Zabolotsky 1993: 450): they are the true animal comrades, liberating figures from the worst communist prison.

* * *

Satantango represents a tired Hungarian village that recalls similar industrial enterprises in the Czech Republic, Ukraine or western Poland. Most of them resulted in their residents' deep social alienation from reality. While a variety of collective and family farm models functioned in different places in Eastern Europe, the literature of this period reflects changes in human–animal–environment relations representative of rural culture across Soviet Eastern Europe. These changing relations are sometimes directly connected with political stagnation, as in the case of *Satantango*. Collectivization of villages created an opportunity for writers to go beyond the anthropocentric framework of literary and cultural expression and listen to the voices of non-humans, even during the period of high Stalinism and deep communism, when the boom in industrialization had contaminated the language, destroyed human agency and polluted culture. It is striking that even weariness can be a great subject for the imagination, where human and non-human witnesses come together in cultural memory and co-create the environmental history of this region.

Part Three

The Earth's Memory

1

Mining Narratives and Their Historical Background

Not only does collectivization continue to cast a shadow over the Eastern European landscape, but so do decades of intensive mining in the former republics and satellite countries of the USSR. Weary cities, towns and villages have been damaged by the extraction of radioactive uranium and by the coal industry's contamination of the soil and pollution of the water. These are only a few of the environmentally harmful, hazardous legacies left on these sites. Thanks to support from local authorities, governments or foreign investors, some of these industrial plants are still operating despite the large amount of carbon dioxide and pollution they emit. The mining industry and related retrograde fossil fuels industries have consequences for ecosystems, health and climate that involve both the legacy of the Soviet past and the continuation of extensive extraction. According to the Polish Central Statistical Office, in 2016 Poland was the fourth-largest producer of hard coal in the world and the third-largest of lignite, whereas Russia was the third-largest producer of hard coal (after the United States and Australia) and the second-largest of brown coal (after Germany) (Polish Central Statistical Office 2017). In the Soviet period, slogans such as 'the country is standing on coal' were popularized in Poland. At the same time, the ideology of communism strengthened naive beliefs in the Soviet bloc and its satellite countries that mining was being developed to benefit national societies, not just Soviet Russia. Even after the fall of the USSR, there remained a strong conviction that mining was a common good – one of those precious treasures of the post-socialist states, a tradition that had to be kept. Nowadays, debates about mining are not only national but also regional. The Soviet heritage is an obstacle for withdrawing from the dirty lignite mining in Germany, since much of brown coal production is concentrated in the regions of the former communist east (The Local.de 2020). People's employment has been dependent on those mines for decades. The same situation involves Poland, the Czech Republic and Romania, who also extract lignite (Appunn 2018). The support for action on climate change in such coal-dependent countries and regions is constricted. Even the health arguments are not enough. During COP24 – held in December 2018 in Katowice, which is the capital of coal mining and heavy industry in Poland and one of the most polluted regions in Europe, with thick deadly smog – the president of Poland told international delegates that the country has secured coal resources for two hundred years and intends to use them.

In this part of Europe, acute scepticism about climate change is one of the reasons why the former Soviet republics and countries of the former Eastern bloc might in their own way represent what Anthony Giddens, influenced by Ulrich Beck's 'risk society' studies (Heise 2008: 146), called 'risk culture' in a Western European context. These post-communist states are not only alienated from the dominant discourse on pollution and environmental damage, but also tend to establish their own narrative regardless of scientific data – overcapitalization in the coal industry is just one of many examples. As Ursula Heise noted, 'the study of risk perceptions and their sociocultural framing must form an integral part of an ecocritical understanding of culture' (2008: 13), but it must also problematize different cultural approaches to the human–environment relationship, especially in societies which are deeply affected by the Soviet mining past.

I will briefly summarize the historical context to situate the mining industry in relation to the environmental culture of Soviet Eastern Europe and its energy crisis discussion. I focus in particular on the communist past of Lower and Upper Silesia, the Polish- and German-speaking parts of the region, as I show that environmental risk narratives are nested in cultural memory through parallel mining narratives about coal and uranium. The history of Silesian mining involves a border that was redrawn by the world wars and post-war treaties as well as political manoeuvring and relocations of people from the Eastern European borderlands. The region's emotional expressions of local identity disturbed by the Soviet intervention need to be studied more to understand its environmental culture.

To sketch out the history of mining in the Soviet bloc countries, it is necessary to start before 1945 when the majority of workers were political prisoners in forced labour camps. This was a period of expansion for such gigantic facilities as Magnitogorsk (Magnetic Mountain), where a vast new metallurgical plant complex was built around iron ore. As one of the steelworkers recollects, when he finished his education in the village school, he heard a rumour that 'the biggest plant in the world would be built at Magnitnaia mountain' (Fitzpatrick 1994: 86).

The construction of Magnitogorsk soon became a theme in Soviet Russian literature of the first Five-Year Plan, with its utopian optimism and belief in a new society built by proletarians devoid of individuality – as in one of the first novels documenting this period, *Time, Forward!* written by Valentin Kataev in 1932 (1933). Magnitogorsk was in fact considered by early Soviet leaders as the most representative enterprise of the 'superior industrial age' to overcome 'Russia's historic "backwardness"' and conquer the steppe (Kotkin 1995: 33) – the landscape of the Russian soul (see Figure 6). Therefore, the development of Soviet mining was not only harmful for the environment, but also intentionally opposed the cultural landscape linked with Russian identity: 'the steppe is freedom, open space, limitless. It is the motherland. And it's also the beginning of the Russian epic' (Vitale interviewing Shklovsky, in Vitale 2012: 112). Mines and other infrastructure for 'conquering the steppe' were also built by severely exploited workers. According to the poet Andrei Voznesensky, the steppe would never again stand for the Russian soul: instead, it represents 'genocide', 'the graveyard of souls', where 'a blizzard of passports whirls over' (1991: 112). This

Figure 6 'Smoke of chimneys is the breath of Soviet Russia.' Early Soviet poster. ©Wikimedia Commons.

tragic legacy of labour camps in the Soviet Union was described by Anne Applebaum in her book *Gulag* where she stresses the lack of commemoration and remembrance of the victims of Soviet repression in the post-communist Russia (2003).

Before 1945, the Soviet Union's mining industry flourished. For example, coal extraction rose from 29 million tonnes in 1913 to 166 million tonnes in 1940; steel production from 4 million to 18 million tonnes; and oil from 9 million to 31 million barrels (Lane 1978: 56). And extraction did not slow down after 1945, as the USSR was enriched by the new Soviet-dependent states and their deposits of hard and brown coal and uranium. East Germany, Czechoslovakia and Poland entered a period of rapid development, ignoring its human and environmental costs. A term was even coined for coal-burning power plants in Soviet East Central Europe – the 'black triangle' – which included plants in south-eastern Germany, south-western Poland, and the western Czech Republic (Bohemia) (McCarthy 2000: 220). Today some parts of the region are still blackened, especially during the winter season, by persistent soot and smog.

The pre-industrial history of mining in Central and Eastern European countries, including Poland, Germany, the Czech Republic, Slovakia and Hungary, began in medieval times. Coal was used in small plants, such as forges, brickyards and distilleries, and was mined on a negligible scale (Jaros 1973: 17). Lignite (brown coal) started to be extracted at the turn of the nineteenth century (Jaros 1973: 22). More intensive mining took place in the second half of the nineteenth century in response to rising demand for coal for railways, industrial plants and electrification. But the breakthrough came after 1945, when Eastern European communist leaders ramped up the mining industry to its largest scale ever in conjunction with *tovarishes* (comrades) from the Soviet Union, who took a lively interest due to the wartime destruction of Soviet mining infrastructure, including in the Donetsk Basin (Jaros 1973: 39). People across the whole Soviet bloc worked on Sundays, holidays and round-the-clock shifts (Jaros 1973: 40) in mines nationalized by communist governments for the 'glory' of the socialist economy (Kopanic 2000: 323).

Work in the mines was organized on the basis of a planned, proportional development of the national economy, in which mining was considered necessary for the development of other areas, such as transport and especially the munitions industry, which comprised 40 per cent of all industrial production in the Soviet Union during the Cold War (McNeill and Engelke 2014: 156). During the Second World War, mining was not destroyed to the same extent as other industries, and was most important to rebuilding the Eastern European countries after 1945. Soon after – during the Six-Year Plan (1950–5) – its development took off at an absurd pace in connection with Stalin's wider economic policy, which led to the industrialization of all kinds of production in the subordinate republics and satellite countries. Another boom occurred in the 1970s, when dozens of new mines were constructed. In Czechoslovakia and Poland, mining was granted huge state subsidies, a privilege primarily due to the specificity of miners as a social group along with their multigenerational traditions of underground work. The industry's special protection persisted until the so-called transformation of the socialist economy into a capitalist one in the 1990s, when demand for coal fell by almost half (Kaczorowski and Gajewski 2008: 201–2).

Mining history and its environmental impact on the communist countries have not been discussed through literature. During the Soviet period miners were either extremely exploited (in gulags localized in the former USSR) – this history of human trauma has been reflected in literature, including Varlam Shalamov's *Kolyma Tales* (1994 [1954–73] and Alexander Solzhenitsyn's *The Gulag Archipelago* (1973) – or they were beneficiaries of the socialist industrial economies (as in Poland). Perhaps, this is the reason why the environmental history of mining has not been ecocritically analysed within the cultural memory of Eastern Europe, despite the existence of literary texts written by both the communist and the post-communist generation of writers.

As we will see in the next chapters, scholarship about literature, history and memory can be used to understand sociocultural perceptions of environmental hazards, vulnerability and resilience. Reconstructing environmental cultures contextualizes and problematizes variants of such perceptions, especially in societies with long mining traditions. After the collapse of communism and the Soviet managed economy, and despite integration with the European Union, which demands higher regulatory standards than other places, Eastern Europe, including the former East Germany, continues to support and subsidize polluting industries such as coal mining. Can the literature of these mining cultures – in which problems of risk and catastrophes constitute cultural memory – participate in contemporary debates about calculations of such risk? What kinds of new voices and testimonies should then be included to make cultural memory function as a valuable and useful environmental discourse? In this part, which I call 'The Earth's Memory', I track literary narratives where the environment, damaged by mining, witnesses geological trauma; where it turns into an ecocentric voice and gets out of the ground to represent the destroyed landscape. This analysis is deeply impelled by the ecocritical perspective, which has the potential to reorient current political and sociological knowledge through literature.

Ecocriticism seeks to interpret how people related to the environment in the past, and then to reform how they relate to it now by means of the environmental imagination. Specifically, in this case, I want to show how the environmental history of particular regions such as the mining areas of Silesia, and of events such as the Chernobyl nuclear catastrophe, to which I refer in the next part of the book, has challenged relations between humans and non-humans. I also aim to demonstrate how the extraction of environmentally harmful materials, such as coal and uranium, has influenced the language used to describe their hazardous nature. Ecocriticism, by reinterpreting and re-evaluating literary sources, disturbs existing patterns of adaptation, which have not been culturally updated, and which 'are basically seen in opposition between people and nature or hierarchically – people over nature' (Barcz, Buchta-Bartodziej and Michalak 2018: 264). Such ecocritical intervention includes close analysis of cultural narratives that represent mining sites and their environmental history, and address changing environments and hazards anew. Ecocritical reading also deconstructs human-oriented relationships with the environments, and anchor vulnerabilities and resilient responses to risks in mining narratives, which may include mountains and particular landscapes as geological victims of the accelerated extraction industry during the Soviet era.

Environmental risks involve more and more different phenomena, which are increasingly directly interconnected, like coal mining with global warming, the nuclear industry with the toxicity of nuclear waste, and so on. In this part of the book, I am interested in asking how cultural examples from Soviet Eastern Europe underscore the risk perceptions we can identify in the present day, and in locating imaginary possibilities to replace the role that mining has played in constructing national cultural discourses in the region. Starting with narratives of coal and uranium extraction, I focus on human intervention in the geological sphere. I extend the notion of human memory to 'the earth's memory' by asking how material damage in the earth's geological layers is brought to the surface in culture, and how the vulnerable, passive earth's voice might be channelled through these narratives.

The terms 'coal' (from the Old English 'col', meaning 'glowing ember' and 'charred remnant') and 'mining' (if derived from the possessive 'mine', Duckert 2015: 238) are entangled with the energy crisis and pose a challenge for possible narratives to address it. These might be categorized as anthropocenic narratives, in the sense of belonging to the Anthropocene, not only because of the geological context but also because of the anthropogenic character of environmental degradation, in which the decisive factor is how people interact with the geological environment, especially in mining regions such as Silesia.

Silesia, as the Eastern European capital of coal, has not yet been incorporated into cultural study of the Anthropocene. However, intensification of extractive industries there during the communist period made mining – including its risks and disasters – one of the topics discussed in national literatures. Therefore, since the end of the Second World War, mining has become not only part of a socialist literary canon that focuses on working-class heroism, but also a constitutive part of cultural memory for this period. For example, the novel *Urodzeni w dymach* ('Born in Smoke') (1965) by Leon Wantuła, who himself initially worked as a miner, charts the decline of mythical coal by foregrounding health and environmental issues. Silesian industrial cities, which were portrayed in the Soviet period as bastions of cultural and technological progress, started to function in literature as unfriendly 'black cities', full of smog and darkness as in the short stories and reportage of Tadeusz Różewicz (1965, 1973).

Such literature – especially post-Soviet examples of it, as we will see in the following chapters – also helped demystify coal, which had signified prosperity, by illustrating its destructive effect on the environment. It put an end to mythization and anaesthetization of coal and mining, especially by Upper Silesian writers like Gustaw Morcinek, who believed that coal was a regional treasure – not Polish or German, but a common Silesian good. However, the Polish communist government, which was organizing massive repatriations of people from the East to the West, began renationalizing Upper Silesia and its coal resources by empowering politically neutral folk culture and legitimating the new borders with the deceptive concept of 'Regained Territories'. By this term, the authorities referred to the remote medieval past when Silesia temporarily belonged to the Polish state (Browarny 2019: 425; 466–8; 474–7). This historical and cultural manipulation idealized post-war 'reconstruction, hard work, upward mobility, civilizational progress' (481) and cut off such mining regions as Silesia from their identity. Repatriates were foreigners there, which influenced their relationship with Silesia's landscape and environment, and shaped later scepticism about the ecological destruction that accelerated mining during the Soviet occupation had caused. Unfortunately, this ideological Polonization was not only the case in 'regained' territories. Nevertheless, Silesian literature strongly represents the huge scale on which cultural memory was destroyed in one Eastern European region, and offers some ecocritical responses which are discussed here under the umbrella term of mining narratives.

Figure 7 Halemba coal mine's shaft with Polish flag. Photo by Krzysztof Duda, Foter.com. CC BY.

Mine narratives especially involve the history of deep mining (see Figure 7): how people put themselves in dangerous environments; how they 'disfigured' the landscapes with mine tailings and slag heaps; and how they 'removed' mountains without any restoration, which led to 'accelerated erosion and occasional landslides' (McNeill and Egelke 2014: 12–13). No less dangerous and destructive for the environment was drilling for oil, which led to many leaks, for example into the Caspian Sea (McNeill and Egelke 2014: 14), or pipeline accidents that affected landscapes and indigenous people in the Arctic Circle (McNeill and Egelke 2014: 20–1). The use of nuclear explosions to locate oil, in turn, caused radioactive pollution in Siberia (McNeill and Egelke 2014: 19). Related controversies include Soviet hydroengineering projects and the disastrous consequences of the so-called 'peaceful' usage of atomic power as manifested in the Chernobyl catastrophe. These, together with today's deadly air pollution because of oil and coal combustion in many of Eastern Europe's cities and suburbs, define mine narratives as cultural interventions that push against rationales for rapidly developing societies in ways that pollute and harm the biosphere in the former Soviet Union and Soviet bloc countries. Therefore, discussions about the Anthropocene and its related concepts, such as the Great Acceleration, the Era of Man and the Capitalocene, which diagnose global environmental threats, provide the background against which particular societies might draw on historical and cultural studies to understand environmental problems better. Though Silesian literature has not yet been studied in this context, it has much to contribute to these debates because of the area's history of intensive mining and because of mining's place in its cultural memory, as evidenced in this literature.

Accordingly, mining narratives entail not only the environmental history of Silesian extraction, but also narrative methods that represent these major environmental threats. They can be viewed as part of the cultural response to the Anthropocene, which non-officially designates the current geological epoch of the Cenozoan era, following the Holocene, and which has significantly reframed contemporary ecocritical theories. On the one hand, the term 'Anthropocene' underscores the weakness and vulnerability of human actors, including their unsuccessful, unending battle with raging and increasingly active environmental disasters that are only multiplying. It also uncovers humans' interdependence with such concrete spheres as terrestrial geology, which normally does not form part of people's everyday experience. However, scholars have pointed out that the anthropogenic era was discussed much earlier in the context of Russian geology. In the 1920s, during the last years of his life, Alexei Petrovich Pavlov described this era as the modern period, when 'in our eyes, the power of man is growing infinitely as a geological force' thanks to cognitive abilities that enabled humanity to take control of the whole biosphere (Angus 2016: 11, 27). The human race had achieved a force and destructive agency equal to that of the elements, and this 'human element' would come to shape the Anthropocene (Raffnsøe 2016: XIII); in other words, 'humans can now be considered an elemental as well as a geological force' (Alaimo 2015: 299).

The problems that arise from human activities such as mining, and literary response through the narratives of mining, contribute to environmental humanities' discussions as in the case of the Anthropocene and human impact on geology. This discussion absorbs the humanities and leads to new questions about their engagement with the catastrophic challenges of global warming

(Clark 2015: 147–54). Or, even more radically, environmental crises are taken as foretelling that the end of the world – as in some references to the Anthropocene – has already appeared on the horizon (Morton 2013: 7). Conclusions are still being drawn and boil down to a repeatedly asserted claim: if we do not stop extracting and using fossil fuels, we will not mitigate the consequences of climate change, but instead accelerate them. Literature, however, digests catastrophic news in its own way and gives voice to a damaged environment in the time of the deepest crisis, as mining narratives show.

Mines and their metaphoric dimensions exemplify cultural potential in reference to widely discussed environmental risks and challenges because they may be understood more broadly than just as sites from which specific deposits useful for humans are extracted. Mines may represent both the poetic process of discovering what is buried and uncovering layers of past geological epochs and what has been laid down in the Anthropocene. To what extent can we express the Anthropocene in a way that merges human and non-human history and culture with the earth, geology and the environment through mining narratives and fiction? How does cultural memory intersect here with environmental memory?

During the communist period, the extraction industry was not only praised in the socialist states, but also figured widely in all kinds of art, both propagandist and not. In literature, mining was already a major theme of Silesian writers before the Second World War. For example, Gustaw Morcinek, who was promoted by Polish communist authorities, wrote a popular children's novella, *Łysek z Pokładu Idy* ('Lysek from the Ida Shaft') (1933), that is still read in schools today. It portrays a horse working in a mine and the terrible conditions for horses working underground that caused them to go blind, and tells a story of friendship between the animal and a miner. This text now can be interpreted as a powerful ecocritical example of humans and animals sharing the same fate in a mine.

However, only the first generation of post-communist writers, born in the late 1970s and early 1980s, recognizes the direct negative environmental consequences of the extensive mining industry and track the damage that affects the residents of post-mining areas. Szczepan Twardoch, in his novel-saga *Drach* (2014), shows how mining shapes layers of history through the example of three generations of coal-mining families settled in Upper Silesia. But, as we will see in the next chapter, the narrator, the omnipresent and omniscient Drach, is the exploited and penetrated earth itself, which observes and comments on the chaotic lives of people from beneath the ground. The history of uranium mining, another major component of Soviet-era industry, figures in Filip Springer's reportage-style work *Miedzianka* (2011, subtitled *The History of Disappearing*, and translated into English in 2017 as *History of a Disappearance: The Story of a Forgotten Polish Town*), which I will analyse in the subsequent chapter. These narratives point to the gaps in human memory that Soviet propaganda and colonization caused and, even more crucially, activate the non-human, material testimony of a hollowed-out environment to complement and supplement what remains missing as people remember the Soviet intervention in Eastern Europe.

Miedzianka is about a Lower Silesian town, known before the Soviet era as Kupferberg, that is rich in copper and a spa destination. The author tells the story of how the 700-year-old town was completely destroyed – together with the mountain on which the city stood – by the Red Army's extraction of uranium ore from it before 1954. Patchy cultural memories of the period of Soviet uranium mining echo and are echoed by the actual gaps and holes that mining left in Miedzianka mountain.

Uranium was the last material excavated in Kupferberg, ending the centuries-old tradition of mining valuable materials such as copper and silver there. Before the war, these traditions shaped the identity of this German-speaking town located in Lower Silesia. The history of Polish uranium extraction is relatively short (1948–68), although mining of it began in the interwar period on the German side of Silesia, which had been divided by the Treaty of Versailles. Tens of tonnes of uranium were taken away from this location, while the Soviets took several hundred tonnes away after the war (Pyssa 2016: 48). These deposits were in Miedzianka, Kowary and Kletno. The demand for uranium rapidly increased during the Cold War, as atomic bombs were built not only in the United States but also in the USSR; consequently, in 1945, all uranium resources in Czechoslovakia and in Germany's Soviet occupation zone were secured by Stalin, who later took over those in Poland, too (René 2017). Miedzianka and other uranium mines were part of the Soviet 'atomic archipelago', along with 'secret cities built for nuclear research, fuel-processing sites, bomb factories, and test sites' (McNeill and Engelke 2014: 163).

Research on uranium had been intensively conducted since the turn of the twentieth century, including by the Polish chemist and physicist working in Paris, Maria Skłodowska-Curie, whose (married) name was used for the unit of radioactivity, the 'curie'.[6] The destructive power of uranium radiation was already known, although not its harmful effects on health and its possible link to cancer. Its radiation was similar to that of radium obtained from uranium ore, which was used even in cosmetics, watch manufacture and healthcare. However, uranium was a more exciting material due to its fissile character, which suggested the possibility of creating both an atomic reactor and a fission bomb. Two weeks before the outbreak of the Second World War, Niels Bohr published an article in which he confirmed that the latter was a feasible target (Klementowski 2010: 22), and the Manhattan Project went on to successfully build the first atomic bombs. Those who died in Hiroshima and Nagasaki were not only victims of the explosion itself. They also died from drowning in the river during rescue attempts, and from burns, radiation sickness and radioactive contamination (Poolos 2008: 97–101, 104). Although the number of people killed directly by atom bombs was far larger than the number killed by the Chernobyl explosion, both events raise questions about the total number of affected people and the scale of the environmental damage.

The determination of the Soviets to obtain uranium intensified, especially when Stalin, through spies, learned that the atomic bomb was incomparably stronger than other types of weapons. Hence the urgency, attractive wages and secrecy of the uranium mining described in *Miedzianka* – for example, a paper mill was officially registered there instead of the mine. Miners' memories, both those reported by Springer and those from other sources, such as the radio play *Miedzianka: Stalin's revenge for uranium* (Radioram.pl 2008), confirm the effects of increased radiation in the workplace, the lack of regular monitoring of radiation levels in the local environment, the lack of employed dosimetrists and, finally, inadequate radiation protection in Soviet uranium mines (Borzęcki 2004). The danger was underestimated because radiation did not kill immediately, as in numerous mining accidents, but later, even up to several years after someone stopped working in the mine.

Both Twardoch's and Springer's narratives link the environmental risk of mining industries with individuals' lives; above all, though, they chronicle the activity of people intervening in the earth's geological layers. They show how the elements of earth resist human control and mastery, especially in the context of late socialist modernization. Drach is the voice of the exhausted earth, which participates in the chaotic, emotionally fraught history of post-coal Silesian land inhabited by people of different ethnic identities (e.g. Polish and German, including non-Silesians). *Miedzianka* – a post-uranium place – evokes an invisible hazard that can be read only through what is preserved in post-industrial ghost towns like the ones described in the book. It fills in the cultural memory of the recent post-war history of extracting this material that is dangerous and traumatic for people and the environment. Narrating these places, which still exist on the map, mobilizes not only human but also environmental memory. Social history is revealed as intertwined with nature, as coal and uranium deposits – artefacts of the Earth's memory over deep, inhuman timescales – play a critical role in Soviet hyper-modernization. Published in the post-Soviet era, these two texts recover traumatic cultural and environmental memories that had been suppressed for more than half a century.

2

Unearthing the Story of Coal: *Drach*

The novel *Drach* involves three generations of mining families who are embedded in the landscape and history of Upper Silesia. Its omnipresent and omniscient narrator, Drach, whose name comes from the Silesian word for 'dragon', tells their story in an exceptional and mysterious non-human voice. His, or its, godlike presence represents the exploited and excavated earth. In the context of Silesia's environmental history and extensive coal-mining industry, Drach's voice combines cultural memory with ecological trauma. The geological poetics textured and vertical in time and space, through which he speaks and expresses his nature, parallel the mining environment where he dwells. Drach's ambiguous position in regard to the history of mining and human destruction puts the novel at the heart of the Anthropocene debate, while this powerful earthly dragon's voice provides ecocentric testimony to Silesia's mining past.

The relationship between geological and environmental elements in the novel, through the anthropomorphic figure of Drach, serves as an example of a narrative in which a geologically wounded environment is a character who participates in narrating the history of a region that lost its identity during communism by bringing an environmental perspective to cultural memory. How the novel accomplishes this is the subject of this chapter. The novel monumentalizes the figure of Drach as a narrator of the linked ecological and human traumas of the Soviet period, but the whole narrative explores the deep discrepancy in Silesian identity between attachment to the environment and awareness of the negative consequences of mining.

We are dealing here with a Silesian novel, but not one that promotes Silesian autonomy (though such political movements do exist). On the one hand, this book shows the chaotic nature of historical events and the problematic issues of belonging and identity in Silesia, which was traumatized and shaken particularly during the communist period. On the other, this novel articulates the voice of the earth, in a metaphorical sense, by making the geological dragon speak. Therefore, Drach is precisely located – as the spirit of Silesian land riddled with mines – as well as universal, in the sense that all Earth is one. Drach's voice tells the story of a multigenerational family as if through geological layers without filtering the stories through the German and Polish nationalist ideologies that tore Silesia apart. The many layers of this story include post-war intervention, when the Soviet Union took over extractive industries in all its newly acquired territories, as well as the intensive Polonization of Silesia during the communist period to accelerate the mining industry. However, Drach does not favour any of the heroes, whose fates turn out to be cursed and poisoned like their very land. His monstrous non-human voice redefines Silesia's phenomenal bond of the material (coal) and cultural (escaping nationalistic politics and reviving regional myth), and fills this space, traumatized by history, with ecological memory.

Drach observes from beneath rather than from above; his voice is not transcendent, but speaks from the level of mining tunnels in the earth. He comments on the chaos of human lives, immanently linked with all of them. He communicates almost as if he is touching human bodies and their most intimate stories. His divine nature pervades the underground world and is narrated through geospatial metaphors. Distinct entities, humans and soil, are united in kinship here through the more general element that Drach represents. They form a relationship based on biological and existential intimacy and not the domination of one of the parties. Although Drach is cruelly non-human, mercilessly narrating the chaotic realm of human affairs, the intimate relationship between people and the earth comes down to an ecocentric mutual dependence that ties together materiality and narrative.

Drach can become a dangerous element like an environmental hazard, since he embodies all dangers connected with mining, but he himself is also subjected to destructive human *techne*, as in the case of the extraction industry. His monstrosity and dragon-like nature may stand for the cursed history of Silesia, tormented by changing political borders, social tumult and its environmental past. His intimate access to humans' lives, through the soil, but also his distanced narration about their lives, contrasts with being affected by them. This makes him seem like the vulnerable voice of the supressed land, which was destroyed by mining and now tries to speak, even if it cannot be heard. The relationship between Drach's geological perspective and his environmental one can be grasped by this tension. Therefore, an environmental critique emerges from Drach's detached, impassive geological perspective on humanity, and in some scenes, Drach becomes less detached and speaks as an earth that can feel wounds.

At the beginning of the novel, in addition to Josef Magnor, the human protagonist, a pig appears (Twardoch 2014: 8; all translations from *Drach* are mine). The way she is introduced – in a scene where pigs are slaughtered and she attempts to escape – blurs the perspectives of human and non-human narratives. The basic biological process of an animal's life, from birth until death, does not differ fundamentally from that of a human being. Drach takes this point of view, adopting a tone close to that of a medieval morality play that acknowledges universal mortality – though death, here, is not subordinated to divine punishment. Drach's voice comes from a different time-space order where earthly, biological realities dominate the short period of human culture – from a *longue durée* continuum calculated in geological eras rather than years, decades or even centuries – and intervenes on a grandiosely non-human, geological scale in the pettiness of the human world. Beyond the instinctive struggle for life, says Drach, 'there is also the wisdom of a pig, the hidden wisdom of a pig that consents to it [being killed]. In her deeper wisdom, hidden from human knowledge, the pig knows that she must return to the ground from which she was born' (8). Then she becomes an industrial commodity and no one pays attention to her autonomy and agency, particularly her capacity to express fear. The voice comments: 'Next generations of pigs turn into human food and then fertilize the earth. Pig slaughter with a knife has almost disappeared' (174). In this case and in others, Drach's perspective seems neutral, extended in time and focused on materiality (pigs' meat will, or already does, fertilize the soil); Drach does not consider the ethical issues accompanying killing as such.

In the novel, the earth dwells in everything and rises above all, even over all-pervasive death; the earth, embodied by Drach, escapes known temporalities and brings a deep-time, geological dimension to the narrative. Drach straightforwardly speaks for the necessary finitude of the above-ground world, which is hopeless both for a pig fattened for slaughter and for humans slaughtered in war. Yet, at the same time, Drach, who distances himself from the external world, finds his own internal, material wounds inscribed in the course of Silesian history and its tragedies for people and the environment. Neither the First nor the Second World War is different – Drach assimilates these catastrophes just as he digests all the other sufferings of the world, absorbing them like the soil absorbs toxic substances, no matter how harmful they are. Drach is imbued with the world, as if its boundless storytelling structure is the only instrument that can receive and decipher a narrative essence, a spoken thread that binds all beings and keeps Drach together: 'everything from this world speaks for itself and says something else, birds, trees, burnt tanks, people and stones speak something and I hear these words, and these are my words' (52). Drach's ontological position is doubled here. Although materially associated with the human world, he takes the perspective of the whole. He is like some divine being when he says: 'I am not someone who understands, I only see' (77). Or in another place: 'I see everything and see everything at the same time' (101). Drach's uncanny geological existence involves a space-time continuum of such long duration that it encompasses human finitude. Drach's geological voice is too vast for human understanding, since he has a deep-time perspective and draws on more-than-human knowledge.

But at the same time, Drach points to a wound in the earth that recalls a living body, which may indicate the moving boundary between what is human and what is earthly, as in the fragment describing the mine when the protagonist, Josef Magnor, goes to work:

> the szola [mine elevator] falls down, into the darkness, into me, shakes with a roar and a rasp of steel, and next to it, the equally full szola glides up; three floors of eight miners glide in me, fall into my body, three floors of eight miners leave my body, some begin their shift, others end, four hundred, five hundred metres in me, they tunnel corridors in me, the lances make holes in my body, goose feathers cleanse them of the remains of my body grated with dust, they load dynamite in me, blow up, they load the output to the cars, pile to the skip, they pull my crumbled body upstairs and burn me in the house and metallurgic furnaces. They stoke me, me, who sees clearly, they stoke the child of the sun.
>
> 60

The narration is conducted here as if from the bowels of the earth that the miners are violently penetrating. The voice of the earth loses its impassiveness as Drach himself embodies the almost feminine experience of a passive, vulnerable body abused by masculine workers, narrating how what he calls 'my crumbled body' is excavated and then destroyed. This process of exploiting the earth undermines the eternal circulation of matter and life, since it ends in mining and burning coal as harm for Drach's body as earth and expressed by him in this dramatic phrase in the first person perspective: 'they stoke me, me, who sees clearly'. Drach calls himself 'the child of the sun' because coal comes from ancient organic matter, compressed dead plants deposited in the earth. When coal is extracted, burned and consumed by people, the indifferent, geological perspective is substituted with Drach's environmental voice that reflects the merciless intervention of the miners, into Drach's body, layer by layer, as if they were hurting a child or raping a woman.

The animated voice of the earth appears to be a vulnerable living thing (132). However, despite its dramatism, Drach's voice remains throughout the novel emotionally inhuman, distanced from the world of human tragedy and elation, insensitive and nihilistic. It is the wounded voice of a victim who does not speak directly about his wounds, a voice that reveals itself only in Silesia, in the mine – a place that mocks evolution, where resources such as fossil fuels serve humanity regardless of the environmental crisis and immediate need to stop extracting and burning coal. Man is the only species that loses survival instinct and Drach knows it. That is why he is impassive towards people and envisions the future from the earth perspective in the apocalyptic way: 'The light in the dark shines and the darkness overwhelms it' (61).

There is nothing mystical in Drach's luminous dark poetics – his warning voice comes from an exploited body of Silesia representing the matter of the earth, a body that recognizes all human users, among whom miners and steelworkers are distinguished from farmers and compared to parasites, because of the intensity of industrial development in this particular region:

> My body differently bears farmers with their soft lines and gradual transitions, and differently the mine's and smelter's people with their right angles, precisely excavated shafts and sidewalks, and ramps, and hollowed in my body galleries like parasites' corridors harassing me, with their metric system and dynamite. The field ends gradually, even if it is cut off by a river or baulk. The shaft ends geometrically, by a straight line like some part of the machine.
>
> 137

Josef Magnor, observed by Drach himself, is the human example of the Silesian paradigmatic role in understanding the environmental potential in this mine narrative. His fate has a complex, potent, dependent relationship with the Silesian land, though his mode of belonging is neither sentimental nor nationalistic or fascistic. His affiliation with the Silesian land involves a strange union with the concrete matter of this ground, which his body recognizes as the dominant element of his identity. As in *The Salt of the Earth* (2019 [1935]) by Józef Wittlin, history senselessly tosses Twardoch's hero around when he takes part in the First World War. Here history possesses greater agency than his Silesian identity. Josef's identity is shaped not only by belonging to a particular region, but also by specific experiences: by the constancy of dirty, cold unpleasant ground, from the muddy trenches, through his descent into the mine, to his death on the surface of the yard (Czyżak 2016: 41). However, his profession is part of his Silesian identity – the miner is a technically proficient man, *homo faber*. And this foreman threatens the earth the most when he uses it in a systematic and precise way.

In the novel, the environmental dimensions of mining are not only revealed through the non-human, geological voice of Drach himself. They are also represented by the mysterious figure of Pindur, an old Silesian sage, whose sayings renounce human separation from the earth, nature and animals. His aphorisms, starting with the thought, 'The tree, man, deer, stone. The same' (Twardoch 2014: 15), appear to testify to and affirm a utopian animate world where things and beings were once related, but are no longer. Pindur is trying to recover this unity by explaining it, in Silesian, to 8-year-old Josef: '*Sŏrnik je to samo co strōm a człek. Podziwej se, bajtel: to je sŏrnik. To sōm my. A to je strōm. Blank to samo, pra? Takie jest to nasze żywobyci na tyj ziymie, bajtel*' ('A deer is the same as a tree and human. Look at it my boy: this is a deer. There we are. And this is a tree. It is the same, isn't it? This is how our life-being on this earth looks, my boy') (16).

Silesian cultural identity has been significantly weakened and undermined by the course of history and Soviet colonization of this land and its natural resources. At the same time, only in the Silesian dialect does the sage express the myth that characterizes this land and its cultural memory in a region of coal where nature is largely obliterated. This is the myth of Silesian identity, which treated work in the mine as indispensable:

> Everything is one. The earth is like a stern drach [dragon]. We trudge over his body, and in the mines we dig his body, which is the pure sun. In the Holy Bible there is written, in Ezekiel, that a pharaoh is a crocodile that lies in the river, right? ... But this is not a crocodile. In Jewish it is a *tannin* and *tannin* is a stern drach. Huge dragon. You know? And somewhere else you find a behemoth, whose legs are like pipes from the bronze and the tail like a stern tree, like a cedar, ok? ... We are him. We are this drach when we trudge over his body but we are also this body, will you manage to understand that? Like a flea that drinks your blood, she is you, right?
>
> 360–1

Here, Pindur recounts this myth via a dual structure: the *drach* is a monster (the crocodile, dragon, behemoth) and, at the same time, humans are the *drach*. Earth is the body that humanity exploits ('in the mines we dig his body', 'we trudge over this drach's body') because coal, juxtaposed with 'the pure sun', is precious like gold; but, in parallel, humans also *are* this earthly, exploited body ('we are also his body'). We are trapped in the paradox of being the victim and the perpetrator at the same time, like 'the flea'. Pindur affirms what Drach communicates in his non-human, nihilistic voice, which represents the geological matter that has been disturbed by humans. In the context of Silesian identity, the sage's voice indicates the ontological bond between this region's mining culture and the environmental realm, as if mining and all its ecological consequences were part of the Silesian soul and its remorse.

In Silesia's coal-based culture, environmental history is tightly linked with the region's ecocentric mythology. However, after a period in pre-Soviet Silesia when coal was as prized as gold, this post-communist novel articulates an ecocritical perspective through Drach, the voice of the earth, and Pindur, the Silesian wise man. They both show how the region's mining past burdens its memory and identity. Perhaps – as in Lowell Duckert's insightful ecocritical essay on coal – people in Silesia can definitively say: 'we are coal, we are earth' (2015: 255). But the novel also universalizes this experience and shows that the environmental perspective – the perspective of the Silesian earth and the nihilistic, cruel impassive narrator, Drach – disregards petty human lives. The individual is the earth's matter, its circulation, its earthly nature, but the human species, which participates in this cyclical process, continually tries to break out, dematerialize and annihilate materiality, forgetting that human life is a small part of a larger whole. The war eventually reminds Silesian people of this, as Drach recounts: 'And now everyone in the ground, they were born from the ground and they returned to the ground, they rose from the mud, they became mud and they would rise again from the mud, because life is very long, not one life but its cycle' (Twardoch 2014: 19). The dissolution of human history into natural history is reflected in Twardoch's chapter titles. The years of events stretch beyond the finite lives of their heroes, resulting in ranks of chaotic but related dates: '1241, 1813, 1866, 1870, 1906, 1914, 1915, 1918' (17). And this is also the story's structural theme – the contingency of the events described. Human events are a disordered jumble of facts connected by chance, but from the perspective of the Earth's life – counted in geological time periods – these facts seem irrelevant and unnecessary.

Why, then, does Drach speak and where does his voice come from? Like other non-human agents, he escapes human cognitive control and represents the fundamental irony of the Anthropocene: we humans are at once its subjects – we caused the environmental crisis – and yet we are unable either to comprehend it or to avert the profound changes it brings. That is why we try to represent it in fiction or invite it to enter, as if literature offered some exceptional cognitive channel. Drach's voice, a signal instance of this capability, is hyper-objective because his range is based on long duration. Similarly, to recognize environmental risk in the Anthropocene requires considering human life in transhuman space-time categories. Unfortunately, everyday things then tend to appear in other, darker colours.

Drach is unreal and mythic, which may also indicate an environmental resistance deeply rooted in Silesian mining culture despite human unwillingness to face the problems. If so, this narrative may be an attempt to locate agency outside the human world, to make the earth speak. Our awareness lags behind the current stage of environmental change, and therefore the less we know about the Anthropocene the more we need aesthetics and fiction to make it real (Morton 2013; Trexler 2015; Clark 2015). The way Drach speaks also makes him a post-human voice, distant from history and knowledge and at once severely critical and impassively moralistic. Or maybe his inhuman voice manages to reorient Silesian cultural memory from the region's human history of ethnic and national conflict to its environmental history of coal extraction. Meanwhile the Silesian sage Pindur seems to speak for a mystic or utopian oneness of humans with the natural world – and these two voices can converge to make this novel ecocentric.

Coal is a paradoxical material because it comes from organic nature – but when it is burned, it pollutes the human and natural world. Drach's voice, like the Carboniferous geological period that has 'passed', represents a type of narrative that does not offer any theory about how to save living nature or people, but has an impact on the story we tell when thinking about the very near future. It is a narrative where the geological past and the history of coal mining seem inseparable, where the human and non-human meet dangerously, and what can be adequately expressed in this Silesian land, experienced and damaged by mining. In the same way, Drach, who observes everything from underground, from the bowels of Silesia, tells a nihilistic tale of the ephemeral world of people, animals and the rest of the matter. We do not know what will happen after the Anthropocene, but its passing has been detected on the horizon, and this Silesian mining dragon is one of its prophets.

3

The Uranium Narrative: History of a Disappearance

This uranium narrative concerns the history of a medieval town and the small peak on which it sits, located among the Sudeten Mountains in Lower Silesia and rich in various ores, which Springer describes in a book entitled *History of a Disappearance* (2017). This book's narrative is completely different from that of Twardoch's *Drach*, an ecocentric attempt to represent Silesian earth memory through a non-human, geological narrator and Silesian folk wisdom.

The town Springer describes is called Miedzianka, named after Miedziana Góra (Copper Mountain), but it was known before the war by the German name of Kupferberg, when it briefly served as a spa resort. The town began to 'disappear' after 1963, when the Soviets stopped extracting uranium there, not only physically but also from memory – hence the book's subtitle, *The story of a forgotten Polish town*. The Soviets tried to erase Miedzianka from the map, like many other similar places that suffered environmental harm, but the earth, in a voice different than Drach's, speaks. Through the holes and tunnels in Copper Mountain's geological structure that was massively violated by mining, the earth uncovers gaps in memory and shows how people have forgotten its environmental destruction; these gaps now form a unique testimonial to the mountain's history and its centuries-long symbiosis with a town that had to vanish.

The disappearance of cities in Soviet East Europe – parallel to the evacuation of Detroit in the United States – is addressed predominantly in research that deals with the political and economic causes of the phenomenon (Krzysztofik and Dymitrow 2015; Wirth, Mali and Fischer 2012). Miedzianka's disappearance originated in over-intensive mining activities, as is the case with many other post-Soviet cities and towns following rapid industrialization in the 1950s through to the 1970s, which radically increased mining output and led to land collapse. However, the whole history of mining there could be a reason why the town's fall had been preceded by incidents of land collapse in the pre-war period, when it was called Kupferberg, and after the Second World War in Miedzianka:

The ground caves in for the first time under the building housing Preus's smithy and Reimann the merchant's. It leaves a crater so large a wagon could fit inside....

One day, the horses ploughing Mr Franzky's field sink into the ground up to their chests and let out such a terrifying howl that those in the immediate vicinity drop what they're doing and race to the field, their faces pale. Only a few are brave enough to come to the animals' aid and rescue them; the rest gaze from a distance at the horses' heads protruding from the earth and the extraordinary, funnel-shaped cavity around them.

Springer 2017: 11

The narrative style that Springer uses in his reportage suggests, from the beginning, that the life of the town took place in the shadow of major or minor catastrophes. The inhabitants of Miedzianka, at least from the nineteenth century on, experienced successive instances of subsidence, which were inscribed in their lives by the dangerous mountain. However, they also managed to establish a kind of symbiotic relationship with the mountain despite the risk, since mining was possible there for many centuries. When it was no longer possible in the second half of the nineteenth century, the town became a resort and the mountain itself was given a period of rest.

The founders of the city 'were brave people', because 'no one faint-hearted could have founded a town in such a spot [on the top of the mountain]. Nor could they have challenged nature so audaciously by digging holes into the stone mountainside, hunting for precious metals in the dark' (12). The first one who started digging is said to have been a semi-legendary Silesian master miner of the twelfth century. The mountain then became famous for tales of its precious deposits. According to the chronicles, silver was mined there in the fourteenth century and then, in the sixteenth century, copper began to be mined, and Dippold von Burghaus, the first professional mining expert, bought Kupferberg. In the middle of the sixteenth century, 'nearly 160 mineshafts and drifts [we]re dug in the mountainside' (14), which seems to be a dizzying number given the modest size of the mountain in comparison with the surrounding peaks. Until the Second World War, the mining history of Kupferberg entailed a series of attempts to extract precious ores from the mountain, which involved both investments and disasters. In 1927, the final shaft, named Adler, as the characteristic architectural element for the German-speaking Silesian town landscape, was closed, '[y]et the town's mining past will not be forgotten. Kupferberg stands at the peak of a mountain riddled with holes. For whole centuries, tunnels have been bored into the interior, which now – unsecured –

The Uranium Narrative: History of a Disappearance 117

are beginning to cave in and endanger what stands above' (29). Subsequent landslides, which affected both humans and animals, form the prelude to the town's uranium history. The Soviet period of uranium extraction put a definite end to the life of this resilient old town that had managed to incorporate catastrophes into its cultural memory and survival over centuries. As Springer informs us in an interview:

> From time to time, something has collapsed, a field has subsided, or a house decayed, but these were not serious damages. The extraction of uranium in 1948–1952 eventually caused this town to really start flying apart. The houses were cracking before their eyes, whole walls were falling off the buildings ... it was smash-and-grab, how Soviet Russians were extracting uranium, i.e. they were following the ledge and not looking at what was above them, they often dug two metres underground and there could be a house standing on the top of them. This house was collapsing, and they did not do anything. Therefore, the town started to disappear.
>
> Springer 2012

Springer uses two protagonists to recount the story of Miedzianka, Georg Franzky and Karl Heinz Friebe, who represent the pre-war and post-war periods respectively. Their stories intersect because the author aims to situate post-war Miedzianka in the town's long-term history and to describe its recent ending. The account of uranium mining, though, is reconstructed through interviews with people who have been living there since at least 1948. To recreate as many contexts as possible for the 'disappearance' of a centuries-old town, the author uses a reportage-investigative style of narrative (Kosterska 2015: 11). At the same time, as the interviews reveal, the story is still not completely comprehensible and, what is more, presents fragmented, sometimes even contradictory information due to Soviet secrecy about uranium mining, which made the whole history of Miedzianka obscure. We know from the interviews that people were not allowed to speak where they worked, but they also hid the troubled past between Poles' and Germans' migration and resettlement after the Second World War. Therefore, what complements people's patchy testimony about Miedzianka from the Soviet period, and what constitutes the cultural memory of risk that still operates there, is the earth's memory that opens through the holes left in the mountain by extensive mining.

Springer does not reflect on people's patchy recollections or enact that patchiness and the gaps in memory intentionally through the form of narrative. He interviews people and reconstructs the history of the town. However, the earth's memory speaks in this book through the descriptions of the holes in the ground, and by surprising people, communicates, trying to preserve the memory about destructive mining against sinking it into its ground. The residents' lack of critical memory towards the Soviet intervention and organizing uranium mining in Miedzianka seems also connected with the benefits they received after 1945. Although being resettled Poles from the east, they received new accommodations in the town, which had been vacated by Germans who were sent to the west. Attracted to news of a new mine and good salaries, they gladly settled in a place where the towers of shafts were mixed up with church steeples (Springer 2017: 121). Soon after, rifts opened up; children played in abandoned mine tunnels, although it was dangerous, and they dug up graves, while adults removed pavement slabs from the cemetery together with other materials useful for new farms (139–40). The disappearance of the town began at this moment, when the obliviousness of new settlers tore the historical tissue of the place. As in the case of the cursed Silesian land in *Drach*, the place, and above all its mountain, becomes part of the uranium narrative in Springer's book, which uncovers empty niches in the memory of what really happened in this place between people and the mountain, and between Poles and Germans.

The Soviet propaganda defined resettled people in a moralistic way as 'pioneers of the Polish way of life, as avengers for centuries of injustice inflicted on the Slavic peoples and tribes at German hands … a Slavic Nation with an ancient and unspoiled culture' (113). But for Poles it was 'Wild West', scary because of remaining German partisans and tempting due to the rumours about treasures and valuables left behind by Germans (114). Right after the Second World War, Poles equated Germans with the Nazis, broke into their houses, even if they were still occupied by them, dug around, set the houses on fire when they could not find the riches (114). In fact, both deported from their countries, nations were harmed by the shifted border and not all Poles wanted to leave their homelands, which belonged to the former eastern borderland of Poland, but in 1945 they became part of the Soviet Union.

The cultural memory of Miedzianka, like its mine-riddled Copper Mountain, is characterized by numerous gaps that the author tries to fill in with information from interviews with its residents, though not only are people's reports incomplete, so are the archives: 'the uranium extraction is shrouded in absolute secrecy' (166). Yet we know for sure that the post-war extraction of uranium and its export to the Soviet Union (in the years 1948–52) was initially extremely attractive to Poles who had lived through the war and wanted to work for good salaries. Most did not mind working in the secret mine operating under the guise of a paper mill (68); they worked in very difficult conditions and did not

The Uranium Narrative: History of a Disappearance 119

know exactly what material they were digging for. They were also unaware of the impact of uranium on their health:

> We knew what they were mining, but no one said it out loud. Anyway, what difference did it make if we knew, since no one told us how harmful it was? They didn't tell us the uranium was killing us, we only found out in the nineties, but by then it was too late because half of us were already six feet under.
>
> 143

The mine was expanded immensely and incomparably faster than during the pre-Soviet period: 'over four years, twenty-five miles of tunnels have been dug. Some speak of a few, some of over a dozen, some of dozens of mining posts ... all the mining drifts of ZPR-1 [Polish acronym for all uranium mines' administration in the region – Ore Production Facilities] put together run a hundred miles long, and if all the mineshafts were combined into one, they would reach a depth of nearly twenty miles' (169). However, safety was not a priority. Pavements were laid above the vein, regardless of the proximity of the subterranean tunnels (170). Some even testified that people were killed, blown up in useless corridors, and that 'dozens of people were buried alive in that mine – they'd disappear overnight, along with their entire families' (144). Springer does not comment on these reports, but more probable is testimony about the health of the miners, who complained of toothache (144), began to drink more and developed coughs (150). Uranium ore also contains radium, which emits radon, an invisible radioactive gas that can cause lung disease, radiation syndrome, tumours, cancer and cellular degeneration (165). The town's morbidity from cancer was higher than elsewhere, especially in the case of young people (304), though this was only revealed much later. When uranium was extracted in Miedzianka, there were no official 'notices about radiation in the pit', no information about 'risks associated with mining uranium' (166). Regulators of ore production plants, despite inspections, did not conduct any research into health risks until 1956. After the mine was closed, concentrations of radon six hundred times higher than acceptable levels were recorded (166).

The information sourced from people is intentionally jumbled up in Springer's book in order to represent the fragmentation and unreliability of cultural memory about this period. It is still not known what is true in the residents' memories and what is not, whether piles of bodies were found there or not, but witnesses indicate how surreal life became because of contact with the radioactive substance. High radiation levels in the mine caused the death of rabbits, which died after two months underground, and made watches stop (145–6). It is rumoured that the Russians knew about the threat uranium posed to human health; some of them helped and warned the Polish workers not to eat meals and drink water in the mine (146). At the same time, they kept the risks a closely-guarded secret (147–8) – the mine has its own secret police station and informants among workers, and according to some commentators Miedzianka in fact functioned like a gulag (Mońko 1995), because once you became a miner there, there was no way back, or some people were disappearing there 'without a trace' (Springer 2017: 147). And contrastive accounts report how human life in Miedzianka flourished in those four uranium years: 'the miners are rich … some buy motorcycles, others renovate their houses' (167). For the first time in the town's history, the population peaks at nearly three thousand people, but 'the word "uranium" is avoided like the plague' (168).

After the deposits were depleted, the Soviets took all the plans, and the mine was closed and flooded in 1952. At that time, the town also began to depopulate. A geologist employed a year later to inventory 'the labyrinth of tunnels' and 'the inactive galleries' (180–1), and to check whether the mountain could be further exploited to extract copper, concluded that over-aggressive uranium mining and digging too many corridors under buildings had put the town in serious danger of subsiding at any moment:

> These tunnels under the town are only held up with makeshift shoring. The posts won't stand the test of time, and the town will eventually collapse. … The mine disappears … The last shafts are already filled in, the drifts are covered up, the scrap is shipped away, and the barracks are demolished. In a few years only the spoil tips, sprouting with weeds, will be the only sign that anyone went below ground in this place.
>
> 182–3

Uranium faded from memory in the 1970s and 1980s when newspapers started to claim that it had never been mined in the town (292), which was a result of the effectiveness of communist censorship. However, patchy human memory cannot fully testify to the damage done to people and the environment by the Soviets' intensive search for uranium in Miedzianka. The testimony also remains in Miedzianka's ground and can revive cultural memory through the description of many accidents, uncovering the material and geological distortion. The gaps in the earth are a memorial to its environmental destruction but could also symbolize the riddled memory of Poles who inhabited the town left by Germans and wanted to forget that Miedzianka was Kupferberg before the Second World War.

Such gaps in post-war memory in Miedzianka are announced by subsequent landslides, until the town will run people off its territory. Once a cherry tree disappeared in one of the families' gardens, and other people in the town went to look but no one commented on the causes of the incident, no one combined it with the mining past:

> The cherry tree used to be there and now it's gone. There's just a hole in the ground with water pouring in. And if you shine a flashlight into the hole, you can make out the top of a little tree with a few cherries no kids had managed yet to pick … their cherry tree is a long way underground.
>
> 212–13

The last residents of Miedzianka only knew that they had to look at the ground while walking (212). Another time, one of the wings of a Renaissance manor, where young people hung out during summer vacations, collapsed into the earth: 'no one notices in the middle of the night, and when the cooks come for provisions in the morning, they freeze at the edge of a huge hole in the ground' (267). There were some plans to secure the ground from the old mine drifts by pumping cement into them, but 'no one knows exactly how many of these drifts there are' (216). This is the main reason for the final displacement of Miedzianka's residents to new blocks of flats in Jelenia Góra. The new estate fulfilled the dream of leaving the 'ghost town where the houses groan and where after heavy rainfall, you have to check before leaving the house whether there's still a scrap of land outside to set your feet on' (223).

The destroyed mountain was finally abandoned by its people, whose post-war history differed from that of the pre-war inhabitants as the history of Miedzianka's mining differed before and after the war. What remains of it is only an empty space in cultural memory, 'just tall grass' (235), 'clumps of greenery and trash by the roadside' (306), 'ruin' (231), a place of oblivion. This uranium narrative, from a human perspective, seems to be unreal and full of gaps, but from an environmental perspective, such gaps materialize the destruction of Miedzianka and make it more real. One of the last fragments of the book is devoted to Karl Heinz Friebe, who comes back to Poland just before the collapse of communism. He changes his name to Zbigniew Antoni Sieroń to sound like a local. His mysterious experience with Miedzianka's mountain caps this history of uranium extraction that could not be fully remembered without uncovering the holes it left in the landscape:

> When darkness fell, someone at the workshop noticed Sieroń was missing. They set out looking for him, in darkness and with fog enveloping the hill, just like every autumn. Finally, they found him. He lay shouting at the bottom of a

> sinkhole, along with a few pieces of fencing. The hole had opened up right next to the road leading to the shop. They tried to get him out, but nowhere nearby had a ladder long enough to reach. So they tied a few ropes together and descended that way. Sieroń was bruised and scared out of his mind, but in one piece. They didn't have an easy time getting him to the surface. When he got back, he excitedly told everyone the earth had tried to devour him alive. For a long time afterward, they joked that he probably deserved it.
>
> No, the ground under Miedzianka won't let itself be forgotten.
>
> 290–1

Were it not for the Second World War and the Soviet invasion, Kupferberg might still be functioning today as a spa resort with a renowned brewery (and there are some attempts to revitalize it). Springer, through a unique form of reportage, shows how Polish-German cultural memory discourse about such shared places as Miedzianka/Kupferberg is entangled with environmental history because 'the ground under Miedzianka won't let itself be forgotten' (291). Every time the heavy rain falls, 'sinkholes and hollows open up' (291). Memory merges here with oblivion and, as a result, the shrunken Kupferberg represents not only the town's tragic fate but also the destruction of the mountain and its human community. Digging in German graves (189–93), cellars and the ground for any treasures or valuables (286–7) by the post-war Polish settlers and looters is similar to mining, poking holes in the mountain's solidity and raking over a history of wounds. All these events converge in memory, constituting a matrix of human and non-human histories that may now be reconsidered and better recalled through the process of merging environmental history with cultural memory.

These Anthropocenic narratives in 'The Earth's Memory' section reflect on both the environmental risks of mining industries and the human actors who intervene in the geological space of the Earth. The dangerous nature of the extracted materials, both coal and uranium, rebounds on the people, who lose their ability to control an industrialized environment that is damaged further during the post-Stalin modernization of the 1970s and 1980s memorialized in these texts.

Twardoch's novel *Drach* represents the earth's voice, tired from long exploitation and participation in the chaotic, emotionally charged, ethnically divisive history of Silesia, while *Miedzianka* recounts an unresolved investigation into how mining a harmful and dangerous element affected its residents and forced them to abandon the town. These places still exist on the map, bearing witness to the entanglement of human and environmental history. These books use imagined non-human perspectives to retell narratives of mining – which, in turn, enables the partial reconstruction of cultural narratives that were suppressed or distorted by 'progressive' communist propaganda. *Drach* draws on this imagined environmental perspective by giving the voice directly to the Silesian earth, while Miedzianka's mountain voice is inscribed into the stories about holes opening up in the ground to remind people about the town's past.

Silesia is exemplary of intensive mining practices during the Iron Curtain period, but its complex history of shifting borders and resettlement also influences human relations with the environment (Barcz, Buchta-Bartodziej and Michalak 2018) and contributes to multidimensional human and non-human memories of it. The Anthropocenic context here involves the undesired agency of materials harmful for the biosphere – coal and uranium – as they come to the surface in these narratives and speak through the language of damaged earth or hollow mountain. The environmental history of mining in Silesia is retold in literature via a poetics of geological and ecological wounds. They are put in dialogue with missing reflection in cultural memory on the troubled history of this multi-ethnic and multinational region during communism.

Springer's report – what he calls the archaeological dimension of his book – as well as the historical and geographical contexts of Twardoch's novel point to yet another tendency of Anthropocene narratives of the Silesian past: a transition from environmental history to fiction. Silesian literature inevitably incorporates the mining catastrophes and risks that constitute the region's cultural memory. Therefore, Silesian literature represents what living with the risks of a catastrophe could mean in the time of the Anthropocene and, moreover, includes a language of non-human voicing and witnessing that prompts us to think not only *about* the earth but also reflect critically on our environmental past *with* the earth (Duckert 2015: 238). Geological nature, like the mountain in Miedzianka or Drach's voice of the earth, does not stand apart from the discussion; rather, it figures as an active participant in human history and reconstructs our memory. Cultural memory of the Soviet past is as contaminated as the environment, due to propaganda, communist censorship and political manipulation, but what was forgotten by humans is brought back to mind by the occupied environments.

History has ceased to be only human, as it shows up in geological matter – as in Twardoch's and Springer's texts – in the form of abandoned drifts and multi-kilometre-long mine corridors, and Silesian history can no longer be divided into the natural and the human as Dipesh Chakrabarty argues. These anthropocenic narratives radicalize definitions of ecocriticism (or environmental criticism) that involve politically engaged studies of relations between literature and the environment. In the case of ecocritical reading, these narratives broaden our experience of the natural world's ontologically fragile and problematic status and of changes in the natural and material environment in the context of dangerous human intervention into its geological tissue.

Part Four

The Persistence of Chernobyl in Cultural Memory

1

Eastern European Risk Narrative: Chernobyl Memorial

The Chernobyl accident has been widely discussed among ecocritics and other scholars in the environmental humanities. I am especially indebted to the way this work addresses the problem of risk via cultural and literary examples. In the case of Chernobyl, Ursula Heise (2008), Rob Nixon (2011) and Molly Wallace (2016) show how the event has become a part of everyday experience and global awareness, and how it has modified perceptions of environmental risk. However, what has not been demonstrated is how deeply the Chernobyl accident permeated Eastern European cultural memory and this region's perception of environmental risk. From a Western perspective, Chernobyl became a figure for Soviet collapse, ecological ignorance and technological backwardness, while, for inhabitants of Eastern European communist countries, Chernobyl was an existential shock that caused transnational ecological trauma in the region. Unlike the politicized Western reception of the accident, this experience of trauma generated an exceptional language of witnessing that includes human and environmental perspectives. The language of this testimony and its challenges of representing human and non-human experience of Chernobyl inform what I understand here by Eastern European risk narrative.

Though the Chernobyl catastrophe is extremely well known and commemorated, not many are aware that, on a warm sunny September day in 1957, an explosion rocked the pleasant, leafy secret nuclear city of Cheliabinsk-40 (today Ozersk). The city was one of ten closed nuclear cities in the USSR, known by acronym ZATO in Russian (closed administrative-territorial formations). This event has recently been ranked the second most serious nuclear accident after Fukushima (Brain 2012: 222) and in detail reconstructed by Kate Brown in her book *Plutopia* (2013: 231–8).

While historians have thoroughly described a variety of nuclear catastrophes, only Chernobyl has become a cultural symbol – a self-referential sign to the entire Soviet nuclear industry and to the world as well. But other incidents occurred that were comparable to Chernobyl in the scale of their pollution. They were not reported to the International Atomic Energy Agency Safety Commission, in violation of global agreements signed by the Soviet Union in 1956 (Medvedev 1992: 264). Nor have they been inscribed in cultural memory as reference points and warnings. These incidents include the 1957 accident in Cheliabinsk-40 described above, known also as the Kyshtym (the name of the industrial complex) nuclear disaster in the Mayak plant (both names are used but 'Mayak' was the secret nuclear facility) located on the slopes of the central Ural Mountains (McNeill and Egelke 2014: 163–5), along with its lingering contamination (Brown 2013: 189–94); a serious fire at the Beloyarsk station (1978–9) (Mannan 2013: 531); the dangerous accidents at the Kursk plant (the reactor blackout in 1980 and radioactive water leakage, not to be confused with the submarine fire in 2000) (Josephson 2005: 290); and the *Atommash* (the nuclear manufacturing facility on the Volga River in Volgodonsk city, which was built too close to the water reservoir and started to sink) in 1983 (Josephson 2005: 104; all accidents, see also Medvedev 1992: 261–70). Why were they forgotten? Because the Soviet authorities' censorship was so extremely effective – for example, to cover up such facts as evacuating the local population of Cheliabinsk within a radius of at least 200 kilometres (Josephson et al. 2013: 142).

In this section, I focus on Chernobyl, rather than these other incidents, not because they were not dangerous to the environment – the similarity between Cheliabinsk's and Chernobyl's disastrous consequences for people and environment is striking – but because they have not permeated cultural memory to represent environmental risk. And to some extent, it was a success of Soviet propaganda that nuclear risk was not reflected in culture before the Chernobyl disaster; it has been locked like the closed city in Cheliabinsk.

The first documents specifying the areas of Ukraine and Belarus affected by the Chernobyl blast were published very late, in 1989, and those for the rest of the region a year later (Feshbach and Friendly 1992: 12). However, scientists like Valery Legasov and Vassili Nesterenko, and military and state authorities such as Mikhail Gorbachev and Volodymyr Shcherbytsky, gave differing accounts of who should have reported the accident and why it took so long to evacuate people. The city of Pripyat, in particular, was heavily contaminated for more than thirty hours before its residents were evacuated, as documentaries by the Discovery Channel and BBC have shown (Johnson 2006; Murphy 2006). The danger of mass poisoning grew because neither direct witnesses nor the plant's managers could believe the scale of the catastrophe: 'the accident wasn't in the plan. The unthinkable has happened' (Medvedev 1992: 52). The plume of radioactive dust spread for thousands of kilometres before the authorities began to react appropriately (Medvedev 1992: 71).

The days just after the blast in the infamous reactor no. 4, when it was impossible to put the fire out, are described by many as akin to a world war or something even worse that cannot be named, since the enemy was invisible. If all the accumulated plutonium and uranium in the four Chernobyl reactors had also exploded, over 100 million people could have been poisoned and Europe would have become uninhabitable. Therefore, the history of the so-called 'Liquidation of the Chernobyl Accident' was inscribed in Ukrainian, Belarusian and Soviet memory as the heroic sacrifice of people who were called liquidators and bio-robots. In Mikhail Gorbachev's first televised speech about the event, on 14 May 1986, he said that 'the Chernobyl disaster deeply affected Soviet people' and failed to mention the non-Soviet parts of Europe, in particular Sweden and Austria, that were affected. He warned that nuclear energy was out of human control and suggested a meeting with US President Ronald Reagan to discuss banning nuclear tests. Eventually, in the eyes of the international public, Chernobyl became a symbol of the former Soviet Union's poor environmental management and backward technology, whereas for Soviet people it was another occasion for displaying superhuman heroism by fighting the ongoing fire and rescuing the rest of the world. For Ukrainians and Belarusians, who were the most affected victims, it was the beginning of the USSR's collapse. Literary reflections on how Chernobyl changed people's experience and became an inevitable part of their cultural memory started to appear soon after the accident and have continued. 'Nothing can be erased, nothing subtracted, nothing canceled, nothing corrected!' writes poet and Chernobyl victim Lyubov Sirota (2003 [1995]). Chernobyl started to be monumentalized in Eastern European memory, leaving an open question: how to fill it with not only human memory.

What is still so striking in recreating the history of the accident is that witnesses keep repeating that the scale of nuclear risk is beyond human imagination. Cultural memory about the traumatic event of Chernobyl proves how the catastrophe critically challenged human knowledge and agency. It also prompted a massive cultural response, which, for example, Tamara Hundorova discusses in her work on post-Chernobyl Ukrainian literature (2014, 2017), arguing that the disaster forever changed Ukrainian literature and cultural consciousness. She shows how Ukrainian postmodernism reused the apocalyptic tropes to capture the human trauma, while she does not consider how this literature addresses the environmental consequences.

Chernobyl was destroyed, in fact, in two stages: the two explosions were followed by decontamination actions in which 'liquidators' attempted to control the radiation by removing the topsoil, killing animals, and cutting down and burying the grass, trees and leaves. Decontamination changed the ontological status of humans and non-humans alike, as if they had lost their reality when they all became nuclear waste: 'the Chernobyl region evolves into a gigantic cemetery, a mass grave in which there no longer appears to be any difference between things and people, between the dead and the living' (Zink 2012: 105).

Even now, radiation in Chernobyl's destroyed reactor no. 4 is unpredictable and harmful, and must be kept under a specially constructed sarcophagus (the first structure lasted about thirty years before radiation escaped, and a second was erected in 2017). The sarcophagus 'must remain intact for far longer even than the Egyptian pyramids' and 'it will have to be rebuilt many times' (Medvedev 1992: 20). However, what has been left inside and how it might affect future generations remains unknown (Medvedev 1992: 73). The scale of such a nuclear accident is really beyond controlled measurements (Medvedev 1992: 77), although some initial calculations can be made based on data, such as medical records of cancer treatment and interviews with people, that is still being collected (Brown 2019). In addition, the problems of storing nuclear material, which cannot be destroyed, occur, and there is the threat that radiation exposure poses for people and ecosystems when radioactive waste is stored in proximity to them (Bruno 2016: 257, 262–3). All in all, nuclear risk is hard to calculate or weigh, because accidents can be hidden or minimized (Brown 2013: 9). And yet, Eastern European experience with accidents like Chernobyl demonstrate that we should not use nuclear power at all. The nuclear legacy of the Soviet regime and the difficulties in accessing information on risk measures, disposition of nuclear waste, health effects and the scale of environmental contamination strengthen this argument.

The invisibility of radiation itself and of accidents that are covered up resonate with what Nixon calls 'slow violence' when he writes about 'a violence that occurs gradually and out of sight, a violence of delayed destruction that is dispersed across time and space' (2011: 2). However, because Chernobyl traumatized memory, it prompted a literary, cultural and environmental response that has definitely made this accident, and nuclear risk in general, visible. Literature and cultural memory of Chernobyl can result in imagining even more how the possibility of nuclear accidents can be terrifying. Chernobyl keeps the anxiety alive when other nuclear accidents, like Cheliabinsk, have been forgotten.

At the same time, this history is traumatic for people directly affected by the radiation, many of whom truly believed in Soviet communism and its imperium. For later commentators, Chernobyl remains one of the worst environmental catastrophes in history, and one which is inscribed in the receding past of the Soviet Union, which brought the final collapse of the regime, and the people who, until the very end, served the state by coping with this tragedy: 'Ecocide in the USSR stems from the force, not the failure, of utopian ambitions' (Feshbach and Friendly 1992: 29). Even as the Soviet myth is fading with time, more and more accounts of the catastrophe's impact on the non-human world, including both domesticated animals and wildlife, keep coming out. While human and ecological factors inevitably intersect in discourse on catastrophes, in Eastern European cultural memory about Chernobyl, these factors seem now inseparable. In other words, the trauma of Chernobyl encompasses the complexity of relations between humans and non-humans, from mother's milk to meat, water pollution to soil contamination, via 'stratified slow violence' (Nixon 2011: 49). Toxicity and memory meet here at both macro- and micro-levels: Chernobyl's remains linger in everything from the Soviet legacy in the former republics and states of the USSR that were affected by the blast to mutated genes in human and non-human bodies.

The immediate victims of Chernobyl's contamination on the ground included rodents, vulnerable plants and pine forests, while other, more resilient leafy species of trees, such as oak and birch, died a year later (Medvedev 1992: 89). At first, the damage to the environment was so inadequately recognized that the press did not report on it. While the forest immediately around Chernobyl came to be called the Red Forest because its contaminated pines changed colour and died within a few days of the explosion, four hundred more hectares of pine forest also died (Medvedev 1992: 89). Radioactive particles carried by rain fell as radioactive dust on Pripyat's marshes. It is assumed that a significant part of the first radioactive explosion was absorbed by the river's vast region of wetlands (Medvedev 1992: 90). Today this area is part of the uninhabitable exclusion zone – 'a curious biological contradiction in the wake of the catastrophe of 1986: abundant wildlife and resurgent vegetation, far more prolific than in surrounding precincts because free from quotidian human actions such as mowing, weeding, paving and hunting – but at the same time less healthy than wildlife and vegetation elsewhere precisely because of the accident' (McNeill and Engelke 2014: 29). The area is also 'rewilded' by animals, some that were introduced (like Przewalski's horses) and others that came back (like grey wolves). Similar examples of rewilded contaminated sites include the Rocky Mountain Arsenal

Wildlife Reserve and the East Ural Nature Reserve, which was created after the Kyshtym accident in 1957 (see Figure 8).

Due to the natural world's vitality, and human absence, Chernobyl's environment was able to regenerate and flourish. The area began to be considered scientifically interesting – an exceptional 'radioactive paradise' devoid of people and owned by nature: 'Chernobyl was showing me a different view of the future. It was a radioactive future, indeed, in which ghost towns and villages stand in tragic testimony to the devastating effects of technology gone awry. But life in the Wormwood Forest was not only persevering, it was flourishing,' writes Mary Mycio after her visit to the Chernobyl exclusion zone (2005: 34). However, beyond this post-nuclear phenomenon of nature's 'win', the unique history of the Chernobyl catastrophe, and the way it has been seared into Eastern European cultural memory, are reminders of how the relationship between humans and the environment fell into despair during an era devoted to building a communist society. The plant's chimney stacks are visually inscribed in global cultural memory as well, representing one of the biggest risks to the environment, and can be theorized as 'the megahazards of the present' (Wallace 2016: 4), deeply penetrating and deconstructing our ontological and epistemological construction of the world. However, the phenomenon of nature's recovery and the rewilding

Figure 8 Eastern Urals State Reserve established in 1966 after the Kyshtym accident. Photograph: Alla Slapovskaya and Alisa Nikulina. Wikimedia Commons.

of post-nuclear sites help us avoid framing memory anthropocentrically and anticipate the ecocentric ways in which post-Chernobyl trauma could be overcome, or at least redirected into creating a new environmental history of regions affected by fallout.

Perhaps the rewilding of post-nuclear sites can play a similar role to those forgotten nuclear disasters, as the one in Cheliabinsk, and show the gaps in cultural memory after Chernobyl. However, what has been analysed so far was that Chernobyl generated a lot of testimony and literature because it was a traumatic event for people. The forgotten events have not prompted testimony yet, though they did harm people and the environment; and rewilding in post-nuclear sites, though environmentally beneficial in some ways, has not affected cultural memory to address positive environmental consequences. While the post-Chernobyl testimony and literary responses that I analyse in this section do not promise to rework trauma this way, they still broaden cultural memory of the catastrophe and show its multidimensionality: how it equally affected humans, animals and different aspects of the environment. Therefore, the Chernobyl accident can be interpreted as an ecological trauma for various kinds of victims and their complex ecologies. The persistence of this event in Eastern European cultural memory – its metaphorical as well as physical blast radius – raises many questions: how are we to represent it or figure out the objective truth about it? What language of voicing and witnessing is used to address the hazardous and tricky nature of this catastrophe? How does it influence the perception of environmental risk in Eastern European regions? For this reason, I identify similarities between the way in which cultural sources memorialize Chernobyl, on the one hand, and the epistemological challenges of environmental risk in the Anthropocene, on the other. I am interested in Chernobyl's traumatic contexts, the difficulty of understanding it, and especially the way Chernobyl is memorialized as not just a human experience. All these factors have influenced how the accident is remembered now, how it has entered post-Chernobyl environmental culture, and how it represents Eastern European risk narrative.

The traumatized language of cultural memory about Chernobyl treated the accident as an epistemological challenge from the beginning; it sought to understand the objective nature of the event – but could not. That is why, on the basis of Timothy Morton's philosophical concept of hyperobjects (2013), I turn to the epistemological challenges of memorializing Chernobyl, because I identify a tension between the need to objectify this catastrophe, to make it super-objective and monumental in cultural memory, and between the affective type of nuclear risk narrative that Chernobyl witnesses and writers employ. Perhaps even, Morton's theory of hyperobjects derives from such testimony, from Chernobyl catastrophism and literature.

The Chernobyl nuclear risk narrative, and its anti-representational but philosophical character, shows how the event shook memory and disrupted the borders between the known and the unknown, the visible and the invisible. The concept of hyperobjects itself is useful here, and it sheds light on cultural memory which tries to address Chernobyl's dangerous and multidimensional physicality – and what I call the hyperobjective nature of nuclear risk narratives. These narratives aim at recognizing the hazardous relations between what exists and objects of a completely new type – hyperobjects of radiation and toxicity. In Eastern European testimony about Chernobyl, there is a struggle between philosophical and literary language, between defensive reactions and expressions of powerlessness – a fight between rationality and imagination, which I trace in the next two chapters.

Chernobyl's cultural and historical background, and efforts to express and rationalize the dramatic incident, are at the centre of contemporary debates on environmental risk, starting with scholarship on the Anthropocene, the Great Acceleration, and disaster and risk narratives (McNeill and Engelke 2014; Cohen and Duckert 2015; Clark 2015; Trexler 2015; Rigby 2015; Wallace 2016; Latour 2017). Morton argues in *Hyperobjects* (2013) that, in the case of catastrophes such as Chernobyl, we are driven by a constant disappointment in the language of human experience, which is unsuitable for addressing these *uncanny* objects that are impossible to cognize. At the same time, Chernobyl witnesses and writers want to tell their stories. They want to represent Chernobyl and understand what happened there, but they stumble into sublime and philosophical questions about representing trauma. Therefore, Chernobyl literature and testimony contribute to a philosophical tradition that seeks to represent trauma and find adequate examples and metaphors for events that are incredible and non-discursive but real – and in addition, environmentally hazardous when people are forced to leave the known concepts of nature behind.

The experience of language as limitless and burdened with the affective nature of trauma closely ties Chernobyl literature to the philosophical potential of poetic language, which squeezes into the tightest fissures of cognition, as in Friedrich Hölderlin's phrase in the poem 'Remembrance': '*Was bleibet aber, stiften die Dichter*' (Hölderlin 1990: 266), which can be rendered as 'But poets alone ordain what abides' or 'But poets establish what remains' (299). Chernobyl made not only writers but also ordinary people into poets. What prevails in their memories is an effort to explain how the objectively inexplicable phenomenon of deadly radiation affected them and their stories.

Thus, their narration not only refers to the Chernobyl catastrophe's hyperobjective nature, but also is metaphorically contaminated by the fallout and its prolonged consequences. When Morton discusses hyperobjects, he includes the example of the Chernobyl nuclear accident. He indicates that we cannot understand these new objects, whose impact on people and the natural environment massively exceeds human scales of time and space. Instead, we still try to narrate this incomprehensibility, searching for parallels, metaphors, poetics, or what Morton describes in Kantian terms as *attunement* to hyperobjects (2013: 30). As a result, literature and witness testimony describes the Chernobyl catastrophe more powerfully than any discursive language.

In parallel to some description of Morton's hyperobjects, without going into detail about his theory, Chernobyl narratives address the epistemological challenges of hyperobjects. After all, witnesses kept repeating that they faced something unknown. Moreover, the 'slow violence' of its consequences, the spread of poisonous radioactivity, endowed Chernobyl with *longue durée*. Some properties of Morton's hyperobjects match descriptions of the Chernobyl fallout, including the most synthetic, 'viscosity', as well as 'non-locality', 'temporal undulation' and 'phasing' (2013: 24). Viscosity means the hyperobject sticks to everything that is somehow connected with it. In the case of deadly radiation, the concept of viscosity can improve our understanding of how people described the fallout as something that they could not get rid of – it stuck to their bodies, clothes and tools. The same goes for 'non-locality': nuclear radiation spread invisibly across countries and continents, and could not be controlled in any sense (38). Radiation is also characterized by 'temporary undulation' and 'phasing', which refer to its unlimited chain of potential impact that changes unevenly in time and space, such as through genetic mutations in organisms. Witnesses describe the experience of Chernobyl fallout and its destructive closeness as radically alienating them from a sense of security in public space, in their home and in their own bodies.

This hyperobjective and affective style of narration is exemplified by 'Chernobyl catastrophism', which shapes Eastern European cultural memory of the accident and recounts its ecological trauma. It is a literary movement, described thoroughly in the lengthy anthology *Po Czarnobylu* (*After Chernobyl*) (2017), published in Polish and comprising literary fragments, academic articles and interviews. Works of 'Chernobyl catastrophism' appeared not only right after the disaster in 1986, but also for many years afterwards, creating a 'polyphony' of voices, narrations and genres about the accident. In fact, the prominent Ukrainian and Belarusian authors Oksana Zabuzhko and Svetlana Alexievich, before they began writing about Chernobyl, were asked why they did not refer to it earlier. Their answers were similar: they did not know 'why', or rather, 'how to write about it, what method to use, what approach to take ... something stopped me, something was tying my hands ... a feeling of mystery ... we had touched on the unknown' (Alexievich 2016: 25); 'I really couldn't write about it. I tried dozens of times and threw it away because everything was "not this" and "not like this"' (Zabuzhko 2017: 39). And their accounts, like so many others, described 'the world of Chernobyl' as a tragedy not only for people, but also for animals and the natural environment. Such texts constructed a powerful model of ecological trauma in Eastern European cultural memory, despite the fact that there were these other major nuclear events in the Soviet period that are not remembered. Chernobyl catastrophism, though, overruled a hierarchizing speciesism in historical and political studies, and drew the attention of countries outside the Soviet bloc to the severe ecological consequences of the USSR's intensive modernization during the Cold War. However, these writers, who also included non-human victims in retelling the catastrophe, were reconstructing the events and distanced themselves from the revival of nature in post-nuclear sites, including Chernobyl.

The following two chapters address the problem of narrating the Chernobyl catastrophe ecocritically and how this irreplaceable part of Eastern European memory and history can be read in the context of Anthropocene and environmental risk discourse. The event influenced a new narrative that provides an alternative to stigmatizing the explosion as a Soviet technological and political failure by focusing instead on very private and intimate experiences of the catastrophe. Such an approach, transgressing this stigma of technological failure in the handling of Chernobyl, is presented by the East German writer Christa Wolf in her novel *Accident: A Day's News* (1987) and by Svetlana Alexievich, a Belarusian reporter, who received the Nobel Prize for Literature in 2015 for her reportage, including *Chernobyl Prayer: A Chronicle of the Future* (1997). Both writers craft literary and cultural memory narratives that blur the boundary between the private and the public, history and memory, the real and the fictional in a world changed by the Chernobyl explosion. Perhaps more importantly here, they go beyond the anthropocentric narrative of the catastrophe and exploit its cosmic and multilevel ecological layers, which, as we see, have also shaped cultural memory of Chernobyl.

2

Contaminated Language: Wolf's *Accident*

Nuclear risk narratives, which address the megahazardous and hyperobjective nature of the Chernobyl catastrophe and reflect what Rob Nixon calls 'slow violence' (2011), are a special kind of cultural testimony. The concept of witnessing Chernobyl can be a puzzling one, since the people and environments affected were dispersed around the globe. It is easier to identify those most harmed by this tragedy, but it blurred direct and indirect witnessing.

In Eastern European literature after Chernobyl, we find attempts to address the problem of poisonous radiation's invisibility in narratives that reveal toxic traces of catastrophe. Chernobyl's radioactive fallout manifests itself in contaminated language; in other words, the viscosity of the disaster takes away writers' means of expression or changes them beyond recognition. Morton's concept of the viscous, non-local and phasing hyperobject and Nixon's dispersed slow violence, uncovered in literary language after Chernobyl, show how this language of memorializing was affected by the accident, how it became a type of nuclear sensor narrative. The nuclear risk in this literature is refracted through a private, individual writing experience, while it also invades reality in an unpredictable way, crossing national borders and challenging common-sense knowledge. The East German writer Christa Wolf's *Accident: A Day's News* (2001 [1987]) provides such an example of intimate language that struggles with the contaminative experience of catastrophe and reflects the limits of expression under what I call narrative pressure of memory. The book was one of the first literary responses to Chernobyl to be published and speak out against the Cold War propaganda for 'atoms for peace' (see Figure 9) – part of a wider Eastern European cultural phenomenon that demonstrated the complex problem of witnessing Chernobyl and its environmental consequences.[7]

Wolf's narrative, inscribed in a kind of diary and a letter, indicates how strange and alienating the experience of the Chernobyl disaster was for individual people and how this radically invaded the language of commemoration. But because it is a novel that takes the form of the narrator's diary and this fictional

Figure 9 'Atom for peace.' USSR stamp: Radioactive Decay as Symbol of Atoms for Peace. Emblem and Pavilion at Expo '67. Wikimedia Commons.

protagonist is an East German writer, like the author herself, it also poses a question: what does it mean for a novel when it plays on the form of witness testimony in this way? In other words, how do literature and cultural memory interact here?

Ursula Heise analyses Wolf's novel in detail in *Sense of Place and Sense of Planet* (2008) when she discusses theorists of risk society, especially Ulrich Beck and his thesis that Chernobyl was a 'paradigmatic accident'. She argues that Wolf's attempt to understand the catastrophe and track its influence is focused on her protagonist's daily activities because 'the realm of local everyday life cannot be separated from that of global science and technology' (185). It is interesting how even gardening, a cherished activity of taking care of nature, can immediately be turned into a 'riskscape' (Heise 2008: 186). No matter where we are, in Wolf's narrative the materiality of contamination becomes real and transgressive, and her literary language sensitively detects this contamination, not only in everyday language that the protagonist, and at the same time the narrator, of the novel uses, but in cultural inadequacy of representing this new environmental reality.

Accident's storyteller is a middle-aged woman who, on the day she hears about the explosion in the Chernobyl reactor on the radio, is waiting for news from a hospital because her brother is undergoing cancer surgery. From the beginning of the novel this strange coincidence of events forcefully links memory and narrative. The direct consequence of the two events – and the indirect consequences of radiation, radioactive isotopes and cancer – permeate the language of the protagonist's internal monologue, most often directed at her brother, and also change the language for natural phenomena and the meaning of the environment. In the context of the news about Chernobyl and invisible radiation risk, the protagonist's rising uncertainty about the brother necessarily turns into a special kind of focus when the perception of changing surroundings cannot be explained only by the narrator's stressful experience of her brother's surgery. This confluence of incidents affects her language, which is unprepared for the human and ecological trauma happening both at once, 'on a day about which' she 'cannot write in the present time' (Wolf 2001: 3). She uses the future perfect form instead to report back how what structured the past experience of the future got colonized by the imposed reality of these events: 'the cherry trees will have been in blossom. I will have avoided thinking, "exploded", the cherry trees have exploded, although only one year later' (3). And she tries to rebel against this language distortion caused by Chernobyl, as an Eastern German writer, who is aware of the fragile nature of free artistic expression under the Soviet regime. However, experienced by the language of propaganda, she can also easily sense when she is censored and defeated: 'That goal in a very distant future toward which all lines had run till now had been blasted away, was smoldering, along with the fissionable material in a nuclear reactor. A rare case …' (4). She therefore writes about it in the form of testimony when the language of personal experience (of the protagonist) rifts into the experience of Chernobyl, when the reactor is still on fire, and the necessity to write it down (the writer).

The narrator works in her garden on that 'beautiful' sunny day, under 'this immaculate blue sky, this incarnation of purity, where the uneasy glances of millions are meeting today' (19). Readers therefore learn when the novel is set in historical time from what she already knows, due to the exact timeline of the news' spread outside the USSR. Moreover, from the beginning she is aware of the suspense of the Chernobyl accident – that the fire inside the reactor core cannot be extinguished at once because the meltdown produces radioactive lava consisting of burning graphite, and that there is a long deadly fight ahead to rescue the world from a real apocalypse (6–7). 'This' day becomes strangely unique, characterized by further information from the radio and a lack of information about her brother. However, while she has mastered the terminology of her brother's disease, she does not know the nuclear jargon that the media uses. It is like the backside, 'the Janus face' of language (82), which makes her think about the Tower of Babel and the 'confusion of tongues', as if the biblical prophecy were now being fulfilled (84). The narrator's experience of the uniqueness of the 'accident', linked with the story of her brother's surgery, dominates and locates itself in her inability to understand words whose common meanings changed on the day of the Chernobyl disaster: 'This was one of those days during which all the signs we have been shown up to now come to mind without my understanding them' (85).

Apart from a few accounts asserting that a deadly dose of radioactivity produces a metallic taste, most testimony about Chernobyl treats measurable radiation as something that cannot be seen or sensed. The narrator thinks about how it would be invisible 'on the branches of the blossoming trees' (3) or 'exploding'. Radiation is a tricky referent, like a hyperobject, that confuses common sense and creates an illusion that people are safe: 'What I don't know won't hurt me' (13). This is the feeling of the protagonist. But as a writer, she has literary skills to testify how her own literary language registers the new hazardous context, how language is unprepared and how it can be 'hurt'. Especially in how she is putting together quotations from German classics like Goethe, Schubert or Brecht (e.g. 'radiant azure', 'nature shining on me'; see Rigby 2007: 125), it detects the new nuclear poetics and makes the narrator obsessively reflect upon it and feel puzzled. This puzzlement informs her response, less as a witness to Chernobyl, more as a writer, who has to framework her memory into a literary narrative and rebel against this new imposed reality. By using the intimate form of internal monologue, though, she is more a witness who is speaking to her brother (being operated) and reporting the news about Chernobyl, including all her fears, irritation and helplessness.

Wolf combines the writer's and the protagonist's witnessing perspective, as if she wanted to say, on the example of this novel, that such traumatic memory, as in the case of the Chernobyl accident, needs a writer to testify how the event disturbed the protagonist's everyday experience, how one anxiety (because of the brother) mixed with another (the news about Chernobyl), but it is the writer who has to preserve this anxiety in cultural memory. Therefore, the narrator sees all her daily activities, all her observations and all the day's events differently, as if all words have suddenly, as a result of the news, acquired an additional semantic dimension connoting nuclear radiation: 'The radiant sky. Now one can't think that anymore, either' (21–2). Her experience of reality is divided into 'before' and 'after' the time of the catastrophe. This is where the protagonist's testimony transforms: when she separates the past from the present, she draws a line directly in cultural memory.

Because Chernobyl's fallout cannot be perceived, the narrator's growing knowledge about what might be happening invisibly stamps her language with trauma: 'What comes of the suffering which we cannot perceive?' she asks (8). Initially annoying recommendations not to eat fruits and vegetables and not to drink milk – which the protagonist seems to disregard and even teases herself about – make her suddenly realize that there are some – 'they' – who 'have even killed our appetite for lettuce and spinach' (20). She wants to ask about 'them', but the only words that come to her mind break the sentence into an incomplete: 'Who, they' (20), as if she has become reconciled with the new conditions in communication, and more and more aware of the materiality of the danger that might also affect her and her family. Her concern about the immediate threat is greater than her concern about its causes, and it overshadows the question of 'they', of those who may be responsible for the accident. In fact she does not distance herself from the political context of the Chernobyl tragedy: instead of giving concrete names of responsible people ('men'), she testifies to how isolated she is in her experience of the danger, how the whole communist Eastern Europe was isolated during the catastrophe, cut off from the real objective news. This inflected cultural memory in the region, even in 1986, still depended on Soviet power.

Nuclear risk narratives acknowledge that the poisonous amount of radiation cannot be detected without special equipment like a Geiger counter: it goes beyond sensual perception, which is what makes it so hazardous and scary. It is the same in Wolf's text; her protagonist is surrounded by something dangerous that is theorized but not empirically experienced, something that is assumed to occur 'here', in her kitchen. She asks herself while slicing bread: 'How and when are nuclids – another word I have just begun to learn – actually stored in kernels of grain?' (8). Nuclids seem like Morton's hyperobjects: they 'are here, right here in my social and experiential space' (2013: 27), and because they are so difficult to communicate about, they make people feel lonely. The protagonist calls her daughter, talks to her neighbours and speaks to her brother, but she faces the accident in solitude – everything happens in her narrative monologue. Wolf constructs her narrator as a fictitious woman who can play the role of witness. That is why it is the monologue of a witness who deliberately did not isolate herself, but testifies through her diary about how the Chernobyl catastrophe permeated her everyday life of a fictitious woman – and therefore anyone's life. Perhaps it is another way of interpreting this interesting interplay between the author – the writer in the novel and the narrator – the fictitious character. Through this strategy, Wolf inscribes her work into a wider text of cultural memory about the Chernobyl explosion to give voice to distortions of unfolding news about the event that are pouring into the character's life at different rates and contaminating her language, but (de)forming the language of literary testimony.

The protagonist listens as furious and helpless scientists contradict themselves, and concludes: 'now all of that was drizzling down upon us together with the carriers of radioactive substances, such as rain' (Wolf 2001: 27). Here, 'that' refers to something more than radioactive fallout: it involves irreversible alterations in language. The neutral names of things have been tainted forever by the context of a nuclear explosion, and they will linger in memory, stamped by a 'viscous' hyperobject. Language is brought into step with what has happened in Chernobyl, as some terms seem inadequate. Her favourite example is 'cloud', which can now also be a 'cloud of radiation'. However, in her experience, irony mixes with fear. The drama of witnessing is based on the fact that this experience cannot be fully described, because what really happened defies cognition and is therefore a foreign element that pollutes the language. Nuclear radiation is invisible, but in Wolf's text it is materialized in language and affects; radiation contaminates other languages, certainly including that of literature, since 'a white cloud', a term used in a popular song, cannot designate purity and innocence in poetry anymore:

> a song from the time when clouds were 'white' and made of poetry and pure, condensed vapor. But now, I thought while peeling the boiled potatoes, it should be interesting to see which poet would be the first to dare sing the praises of a white cloud. An invisible cloud of a completely different substance had seized the attention of our feelings – completely different feelings. And, I thought once again with that dark, malicious glee, it has knocked the white cloud of poetry into archives. It has, in the space of a day, broken that and almost every other spell.
>
> 55

The cloud of radiation – this 'invisible cloud of a completely different substance' – was chased by scientists and Soviet aircraft trying to predict its direction and speed, but the most radioactive part of it, released during the first two explosions, had already travelled beyond their monitoring. It seems that at first they did not realize how big the plume was, until 7 a.m. on Monday 28 April, when scientists in Sweden sounded the alarm. It is likely that the Swedes would have done so earlier, but it was the weekend and the cloud had drifted freely across the borders of different countries until Monday morning (Medvedev 1992: 194–5).

Meanwhile, the reactor was still on fire and there was a prolonged danger it would explode and massively contaminate the rest of the world. In fact, Wolf's protagonist wakes up at 7 a.m., too (Wolf 2001: 4), and she hears the news about the problems with extinguishing the fire. The words from the beginning of the novel suggest that her story is linked to the timeline of events at Chernobyl. She wants to remember how the trees 'exploded' but not now, not in this moment when explosion means the real possibility of approaching catastrophe. This day is real and, at the same time, unreal because not everything that happens can be confirmed: on the radio the specialists speculate whether the core of the reactor will melt down or not (6). It is an 'uncanny' experience of the reality that nuclear power has got out of control – this uncanniness comes from the intensity of the reported events and their intersections with the protagonist's life, which have to be memorialized in this historically condensed time. That is why, the private and public meet in the protagonist's traumatic narrative, which combines her response to the catastrophe's unbelievable consequences with reflections on how language has been transformed, her alertness and insecurity during her brother's operation and her distrust of medical technology. Her powerlessness in relation to these events is fundamental to the diary's story and this helplessness is also what prompts Chernobyl victims and witnesses to speak. But because the explosion presented such a new, frightening and dire threat, it was politically empowering in a way. Feeling ultimately vulnerable, people in communist countries regained language for their experience, pushing back against Soviet propaganda and the communist authorities' prolonged silence about the catastrophic consequences of the fallout.

The Eastern European experience of Chernobyl, to which Wolf's novel attests, includes the key dimension of living in a 'utopian' political system with a specially constructed language where nuclear atoms were associated with peace. Therefore, the narrator feels anger against 'those men in pursuit of the peaceful atom' who 'were being spurred on by a utopia: enough energy for all and for all time' (30), or when she sarcastically comments: 'We don't even need a war. We manage to blow ourselves up in times of peace' (35). In fact, Soviet nuclear scientists were proud of their pioneering role in the 'peaceful' use of nuclear energy to produce electricity, since the United States and the United Kingdom had, until that point, only used it to build atomic bombs (Medvedev 1992: 227), in 'the deadly American style' (Brown 2013: 232). Soviet propaganda about the peaceful application of nuclear energy was hypocritical, though, because 'the technology was far too expensive for civilian use only' (Medvedev 1992: 226). While, on the global level, fossil fuels seem to be more hazardous and harmful for society and the environment, the way in which Chernobyl lingers in Eastern European cultural memory and haunts its people shows that this nuclear accident left a deep traumatic wound. Especially in anti-Russian Ukraine, distrust of the atom and the Soviet past in cultural memory means that the nuclear industry as a whole is a reminder of the hypocrisy of learning to live with apocalypse. This applies to the 'whole [of] civilization, contaminated by a "melancholic" waiting for its own end like in the case of an organism exposed to radioactive radiation' (Zabuzhko 2017: 38). However, in post-communist countries people still tolerate nuclear power despite their distrust of it because of its Soviet past and Chernobyl, and cultural memory of it in Eastern Europe shapes discourse about nuclear risk there and results in powerful literary responses, such as Wolf's, placing the writer and her reflection on literature in the rebellious anti-regime position.

Language is dramatically unprepared for Chernobyl's blast, especially literature about nature: 'what to do with the libraries full of nature poems' (Wolf 2001: 37). Language cracks disturbingly, which the narrator wants to turn into a joke – 'Shakespeare and Greek tragedy wouldn't do a thing for me now' (18) – but it is black humour for her to discover that the language she has mastered is inadequate to describe this event. The 'spell', which makes literature an aesthetic experience and reconnects with nature, is removed only after a day, this day when the protagonist heard the news about Chernobyl. Immediately, she loses her literary freedom to use language and its connotations, as if she has been deterritorialized from the infinite possibilities of language as art. She loses her orientation because of 'objects of the most wretched kind' (60). All Chernobyl witnesses seem to repeat that 'after Chernobyl nothing can ever be the same'. Wolf notices how the event affected literary responses at the linguistic level: such words as 'radiation', 'glow', 'cloud' or 'mushroom' have been contaminated by this disaster in Eastern European culture and memory but she is not interested in creating a new nuclear poetics. Through her ecocritical approach to literary language, and as a Chernobyl witness, she testifies how scary and total this whole event was, when it spread not only into the most precious culture, but to the spoken language as well.

In Wolf's narration, there is a clear sense of that when she speaks about rain (Wolf 2001: 38–9). In those days, people were scared of the rain, meaning the radioactive fallout, but fortunately East Germany was spared thanks to its dry weather, in contrast to rains in Poland, southern Germany and Austria (Medvedev 1992: 202–7). Something real and apparently safe like the rain was immediately cast out of the created world, as if it belonged to some 'foreign, unknown God' (Wolf 2001: 39). Rain blurs with radiation and fills the speaker with constant anxiety about contamination, which is additionally fuelled by lack of certainty and irritation. Lawrence Buell also describes 'toxicity as a discourse' this way: as a 'discourse of allegation rather than proof' that 'rests on anxieties' (Buell 1998: 659).

Wolf's narration discloses how language is exposed to hyperobjective contamination during a day filled with news about the Chernobyl catastrophe, when some unnamed 'they' – however, we know who they were – got 'the clouds all mixed up' (Wolf 2001: 41). Here, language behaves as a cultural radar, a tool of testimony trapped by its own limits, unable to adequately express the rapid ontological changes of radiated reality. The context of possible extermination, of intertwined human and ecological traumas, return us to questions about language stored in cultural memory and new language that must testify to its contamination.

Thus, the narrator cannot 'switch off' her 'faculty of imagination' (59) because the Chernobyl catastrophe challenges imagination to fight for adequate means of expressing it. This imaginative struggle constantly seeks adequate ways to understand and remember how danger spreads in spite of our knowledge and to locate what Joseph Conrad in *Heart of Darkness* called the 'blind spot', since 'the center of greatest desire has to be located in close proximity to that darkest point', to 'the peak alongside the crater' (89–90). The language of environmental risk in the novel is presented in its 'true inadequacy' (94) to the event, which can be neither properly described nor experienced through the long German tradition of nature writing. However, memory itself creates a narrative pressure that ties the protagonist's experience of listening to the news about Chernobyl with her own personal worries about her brother's surgery. It is a narrative that reveals a special niche in Eastern European environmental culture that Chernobyl has persistently occupied, penetrated and transformed.

3

The Bees Knew: Alexievich's Chronicle

Oksana Zabuzhko writes that, in Kiev during the Chernobyl incident, the bees were the best dosimeters:

> Every day a new pile of hairy corpses darkened on my windowsill, and I was counting: three, six, eight ... After 'eleven' the bees' mortality curve started to drop. The feeling that accompanied me every morning as I opened the window and swept them out of the windowsill with a wet cloth was nothing new for me: I was burning with a sense of species guilt, the guilt of the stronger against the weaker ... Because they did not understand, and I did, I was ashamed in front of all of them, those alive, dead and unborn (Plato would say – I was ashamed in front of 'the idea of the bee'). As if I myself had built this nuclear plant for them, stuffed it with enriched plutonium and blown it up; as if I was responsible for their apian death.
>
> Zabuzhko 2017: 40–1, my translation

At the same time, in this fragment, she describes how the insects' deaths 'registered' the catastrophe and notes that the vulnerability of a creature so tightly interconnected with the human world as the honeybee underlines both the breadth and the intimate immediacy of the disaster, which became a more-than-human ecological trauma.

In *Chernobyl Prayer: A Chronicle of the Future* (2016 [1997]), Alexievich keeps coming back to how people 'betrayed' nature during and after the disaster. For her, as for Zabuzhko, Chernobyl's effect on other species and the environment is a profound moral issue, as if the accident had happened in two dimensions: in a social one for the Soviet Union and humanity, and in a cosmic one for the world beyond humans. However, nature's role in the Chernobyl narrative is missing from dominant histories, so Alexievich tries to redress the balance and inscribe it into Eastern European cultural memory. She cannot interview non-humans, as she does Chernobyl witnesses and survivors, but it turns out that what happened in Chernobyl cannot be abstracted from the non-human world. Animals and nature are integral parts of her interlocutors' stories, as in the case of a beekeeper who is a skilled translator of the insects' reaction:

> In the morning, I went out into the garden and something was missing, the usual sound was gone. Couldn't hear a single bee – not one! Eh? What was that about? And they wouldn't fly out the second day. Nor the third. Later, they told us there was an accident at the power plant, which wasn't far off. But for good while we didn't know. The bees knew, but we didn't.
>
> Alexievich 2016: 31

Other witnesses repeat similar memories about the sensitivity of bees to radiation. The bees 'could feel it right away' and were 'sitting in the hives', 'waiting it out' because 'their nervous system is cleverer than ours' (62). People trusted the bees: 'if the bees are busy, it's still clean' (98). In differing contexts, the bees provided warning or reassurance.

Other omens indicated the scale of the environmental catastrophe. The anglers could not find 'ordinary' worms for fishing because 'they buried themselves deep in the ground' (32); suffocated moles were lying in orchards and vegetable plots; cockchafers and other grubs had disappeared (61). According to a taxi driver, 'the birds were acting as though they were blind, dropping down on his windscreen, crashing into it' (98). 'A stork would run and run across the field, trying to take off but not able to. A sparrow would scuttle over the ground, jumping and jumping, but not able to fly, not able to fly over the fence' (215). Many different species just vanished, heightening people's anxiety.

Throughout *Chernobyl Prayer*, Alexievich describes the catastrophe without any protective shield. Her form of reportage relies on the 'power of documentary prose' (Zink 2012: 101). She captures her interlocutors' stories as both monologues and collective contributions or choruses. Survivors, witnesses or their family members recount their memories; they speak about their experiences with their own degree of drama, at their own pace and with their own selection of examples. Alexievich follows their stories as if she were the only one listening to them, without interrogating or interrupting them, but keeping records, such as when someone asks her to write down something particularly important. Andrea Zink writes that 'Aleksievich brings us face to face with monological words. She does not add anything to the words of the people concerned, and the readers are also left aghast or at least astonished in the face of these confessions' (2012: 103). In *Chernobyl Prayer*, the problem of *mimesis* (Lindbladh 2008b: 41) and the question of how to represent trauma are resolved through a polyphonic form of speech, as if suffering were shared but impossible to objectify. The polyphonic form, as Johanna Lindbladh notes, was identified in literary studies by Mikhail Bakhtin in his work on Dostoevsky's novels as a 'never ending dialogue' (2008b: 50). According to Bakhtin's 'definition of an epistemology which is polyphonic – a structure of knowledge in which many parallel voices do not become subordinated to one monologic, narrative voice', Alexievich's book of confessions is truly polyphonic (Lindbladh 2008b: 51). Indeed, she develops an ecocentric polyphony when she includes nature's 'suffering' and its neglected voice as an inevitable part of testimony about the Chernobyl catastrophe.

Another aspect of this polyphonic perspective is, according to Zink, that such a 'multiperspectival viewpoint ... gives emphasis to the intensity of the moment and its inevitability. Hence the events ... are all presented from different angles' (2012: 102). Human voices testify to how reality has been distorted in the long aftermath of the catastrophe and how the disaster cannot be rationally understood, since it precedes their experience but does not condition it. Therefore, people have different, contrasting opinions about what has happened. Their accounts stretch and deform reality, break their narration and divide their consciousness; they laugh, ironize and cry, since they have been affected by a tragedy that was inexplicable at first and then by silence and misinformation from the Soviet authorities. Some are convinced that the explosion was sabotage, that it was a deliberate act of foreign powers (Alexievich 2016: 11, 188), as if to revive the language of Stalinism (137). Some compare it to a war (177, 184, 200) and say that they had to defend their Motherland (181). Others suggest it was 'a cosmic experiment' (237) or a fulfilled prophecy from the Book of Revelation (the star there is named 'Wormwood', which means Chernobyl), 'a sign for us' (74). Still others say that there's no radiation (43) – 'this threat here, I don't feel it. I don't see it. It's nowhere in my memory' (73) – or that the radiation is not as powerful as has been stated: it pales in comparison to war (47). Their voices create a chaotic web, which is evidence not of the desultoriness of their state of mind but a manifestation of the event's uncanniness. *Chernobyl Prayer*'s environmental response to the nuclear fallout's effect on bees and other creatures sensitive to contamination makes it a powerful narrative that gives nature and people equal voice.

The insufficient information authorities provide to protect people and the environment provokes various interpretations. A politician says that they were proud of living in the 'Atomic Age' and at first nobody was scared; his knowledge about radiation was minimal, and he had not heard about 'caesium in milk' or 'strontium' (Alexievich 2016: 246). He also admits that politicians did not believe what had happened; they thought that this was just 'another accident' (244) and they 'didn't understand that physics really exists' (245). Kate Brown, whose critical research on Soviet and American nuclear disasters is internationally established, notes in *Plutopia* that 'knowledge about radiation was classified in the Chernobyl zone ... knowledge of risk was a closely guarded secret ... the Chernobyl community had no idea about accidents in the Urals' (2013: 284), which means they did not know about the earlier Kyshtym accident, for instance. Such ignorance was compounded by propagandistic representations of the peaceful atom, as Alexievich writes:

> we were men and women of our times who believed, as we had been taught, that Soviet nuclear power stations were the most reliable in the world: so reliable, you could even build one on Red Square. Military nuclear power meant Hiroshima and Nagasaki, whereas peaceful nuclear power meant an electric light in every home. Nobody had guessed yet that military and peaceful nuclear power were in fact twins.
>
> 2016: 27

The liquidators – soldiers and reserve officers who were directly involved in the most dangerous aspects of the clean-up – testify that they often did risky things because they did not care (Alexievich 2016: 81). They identified with the myth of the brave Soviet man (85, 149, 267), with 'Soviet heroism' (268), with 'a culture of superhuman feats and sacrificial victims' (173), with the 'Russian soul' (179) and with the 'Slav mentality' (227) – only to be disabled if not killed or even turned into nuclear waste.

For many others, though, Chernobyl was a traumatic event incomparable to other tragedies and they could not find any reason to feel reassured. This is evidenced by their accounts of intense sensations in their bodies and minds that recall those linked with affective disorders, though they were prompted by consciously processed information. Such witnesses became lost in what they knew and their interpretations of the accident. Their sense of reality just fell apart. Though they persistently tried to describe the Chernobyl disaster and find meaningful comparisons, their reservoir of cultural memory and public knowledge was insufficient because of the totality of this catastrophe and how it permeated their intimate lives. They had lived for decades with the hazards of the Cold War military competition between the United States and their Soviet state, but the Chernobyl catastrophe's impact seemed to go beyond this framework of possible events. It was a new type of war – 'nuclear war', 'a war that was a mystery to us; where there was no telling what was dangerous and what wasn't, what to fear and what not to fear' (84); 'we felt as powerless as prisoners, and at the same time there was fear. And mystery' (101); 'we were preparing for war, nuclear war, building nuclear shelters. We wanted to hide from the fallout as if it was shrapnel from a shell. But it's everywhere, in the bread, the salt ... We breathe radiation, we eat radiation' (133). Chernobyl was eventually identified as a new threat: 'it was simply beyond me. Beyond knowledge, beyond all the books I'd read in a lifetime ... Some completely unfathomable thing was destroying the whole of my previous world' (136). The disaster changed their common, everyday experience, throwing them off balance: 'I am afraid of rain. That's what Chernobyl means. I'm afraid of snow, of forests, of clouds. Of the wind ... Yes! Where's it blowing from? What's it bringing? That's not an abstraction, not a rational consideration, but my personal feeling' (209); 'what had happened was something we didn't know about. A different kind of fear. This was something you couldn't hear or see. It had no smell, no colour, and it changed us physically and mentally' (239). As in Ulrich Beck's observation about the Chernobyl explosion in *World at Risk* (2009), which examines 'life in world risk society', people were expropriated of the senses (116), not only empirically, but also socially and culturally. They lost 'common sense, as an anthropological precondition of self-conscious life and judgment' (116). They were deterritorialized from their commonality, becoming embodied witnesses of Chernobyl's 'slow violence' (Nixon 2011). At the same time, Alexievich notices that, in their testimony, people began to change their language and to speak 'in new idioms' (2016: 26) – like the writer-protagonist in Christa Wolf's *Accident*. Chernobyl witnesses and storytellers radically transformed their language to accommodate the disaster in cultural memory.

For the peasant community of *Polessya* (in Ukrainian) or *Polesye* (in Polish and Belarusian), with their centuries-old folk traditions, Chernobyl was a catastrophe for the rural world, a clash with peasants' 'own relationship with nature, a trusting, not predatory, attitude' (208). They did not understand why soldiers were digging up their vegetable patches, their 'ordinary vegetables', as if it was 'the end of the world' (100). That is why it was so difficult to evacuate and eventually resettle peasants from this historical region; they were deeply rooted there, and some of them, especially old people, stayed: 'They couldn't believe their world had been turned upside down in a single day and that now they were living in a different world: the world of Chernobyl. They had no intention of going anywhere else' (227). Life went on under the radioactive cloud. People were preparing for Easter and laughing at scientists who were talking about radiation (256–7) because they did not know about the scale of the contamination: 'A young woman was sitting on a bench by her house, breastfeeding. We tested her breast milk and it was radioactive. The Madonna of Chernobyl' (202). In Belarus, villages were evacuated but sowing continued. The state's collective farming model had its plan of production and stuck to it, despite the fact that its food was no longer edible (201–2, 261).

The ignorance and inertia were not to last, however. The Chernobyl disaster, and the crisis it represented for the political and ethical system, prompted Ukrainian and Belarusian intellectuals to suggest that it had initiated major political, social and cultural changes in their communist countries. For example, Belarusians, until Chernobyl, were seen as people of the earth, a 'monocultural' agricultural society, but Chernobyl drove their cultural transformation. As Alexievich notes, the accident started 'sculpting something out of us' (243), as if it made this rural society more reflective, philosophical and critical of the imposed political regime.

Tamara Hundorova argues that Alexievich's approach to the Chernobyl catastrophe offers a framework for testimony and memory (2017: 55), and that she shows how this unspeakable experience, impossible to express, brought its victims a new sense of community. This is confirmed by the many accounts of people identifying themselves as 'Chernobyl people' and the Chernobyl experience as 'their memory': 'just like that, you've turned into a Chernobyl person' (Alexievich 2016: 44); 'the world has split in two: there's us, the people of Chernobyl, and you, everyone else ... We don't make a point of this "I'm Belarusian", "I'm Ukrainian", "I'm Russian" ... We're from Chernobyl. I'm a Chernobyl person. As if we're some sort of separate people. A new nation ...' (136). This traumatic identification was also driven by the authorities' failure to recognize them as victims and sufferers. The problem of seeking satisfactory compensation brought Chernobyl people to 'create novel forms of biological citizenship' (Nixon 2011: 47). Their identification with each other was even strengthened by feeling stigmatized outside 'their community' (Alexievich 2016: 193, 231–2). That is why they constructed a narrative about their experience that insists it cannot be compared to anything that they had gone through before – for them it was unique:

> People are always comparing it to the war. War, though, you can understand. My father told me about the war, and I've read books about it. But this? All that is left of our village is three graveyards: one has people lying in it, the old graveyard; the second has all the cats and dogs we left behind, which were shot; the third has our homes. They buried even our houses.
>
> 182

Alexievich writes that this catastrophe 'has surpassed the camps of Auschwitz and Kolyma. It has gone beyond the Holocaust. It proposes finitude. It leads to a dead-end' (2016: 31). This may sound like an exaggeration, but by comparing Chernobyl to genocidal events, Alexievich highlights the combination of deadly consequences for human and environmental communities, the experience of ultimate erasure in complex human and non-human ecologies, and the continuing threat that they will never get rid of the poisonous radiation – that it will stay with them as a Chernobyl effect and stigmatize them forever. Chernobyl cannot be compared to Hiroshima either: her interlocutors point out that there was no atomic bomb, no mushroom cloud, nothing that they knew of before: 'we heard rumours that the fire was unearthly, not even fire but light. A glimmering. A radiance. Not blue, but a translucent azure. And without smoke' (236). These science fiction-like imaginings of the reactor fire, based on lack of first-hand knowledge and anxiety about imperceptible hazards, suggest a narrative of sublimation (Hundorova 2017: 56) that expresses an awkward beauty. A man describes his first trip to the exclusion zone and its disconcerting pastoral appearance:

> I went there thinking it would all be covered in grey ash, in black soot, like in Bryullov's painting *The Last Day of Pompeii*. But I got there and everything was beautiful. Breathtakingly beautiful! Meadows in flowers, the gentle spring green of the forests.... Everything is coming to life. Flourishing, singing ... What struck me most was the combination of beauty and fear. Fear could no longer be separated from beauty or beauty from fear. Everything was turned on its head, topsy-turvy.
>
> Alexievich 2016: 146

The closed Zone appears in testimonies as a special place, with an aura, 'grandeur' and a kind of horrific beauty (103); it 'pulls you in, like a magnet' (105). As a model of a beautiful but damaged world, it was drowned in silence, without the sounds of nature, without even mosquitoes (251). In descriptions recalling Andrei Tarkovsky's *Stalker* (1979), the Zone is represented as absolutely inhuman, metaphysically mysterious but alive. Its ironically rich ecosystem is now protected by the Belarusian state as the Polesie State Radioecological Reserve. As Professor Lidia Tsvirko writes on the reserve's website, the 'removal of any anthropogenic load' allows 'safe existence even for species with large individual areas. The reserve's contiguity to large areas of the Ukrainian exclusion zone gives opportunities for creating a major cross-border nature protective wildlife refuge' (Zapovednik.by).

In fact, Chernobyl changed many witnesses' perceptions of nature. Those who live near the Zone or enter it find themselves closer to non-humans: a returnee says she does not shoo magpies away anymore because 'we're all suffering from the same trouble these days' (Alexievich 2016: 39). Another woman says that 'now I can walk through the forest alone ... I can't feel afraid of the earth, the water. It's man I'm afraid of' (70); yet another returnee says that he started to watch ants because he had 'never really noticed them before' (136). Someone who had just arrived, attracted by the news that the area was reopened, admits that 'there's total freedom here. I'd say it's heaven. There is nobody here, just wild animals wandering around. I live among animals and birds. Who can say I'm alone?' (73).

Alexievich's book enacts what Hundorova calls 'Chernobyl catastrophism' – a powerful aesthetic model – by emphasizing the non-representational character of Chernobyl, contradictory opinions about what happened, radiation's invisibility and the incomprehensible power of its scope to poison (2017: 61). Catastrophism construes reality aesthetically and substitutes phantasms for it (Hundorova 2017: 61), since Chernobyl reality is determined by the limits of representation.

The language of literature and testimony about Chernobyl was concerned with the disaster's epistemological challenges long before Morton wrote *Hyperobjects*. Morton's concept of hyperobjects derives from such testimony, from Chernobyl catastrophism and similar literature. Morton describes 'the overall aesthetic "feel"' (2013: 22) when a sense of asymmetry between the infinite powers of cognition and the infinite being of things is experienced as something that makes the Chernobyl catastrophe sublime and unique. Thus, the catastrophe itself has come to be remembered as an aesthetic object: the colour of the glowing reactor appeared as 'bright, raspberry red ... it was an incredible colour. Not an ordinary fire, but a kind of shining. Very pretty ... we had no idea death could look so pretty' (Alexievich 2016: 191–2). People began developing metaphysical theories about it (Alexievich 2016: 237). They felt a need to philosophize about Chernobyl (Alexievich 2016: 171), and so did Morton, many years after Chernobyl, in line with the preceding catastrophist discourse about it.

The most dramatic accounts that Alexievich quotes are those of women who lost their closest ones, their beloved husbands who served in the first teams of liquidators, as well as the testimonies of the liquidators who survived but witnessed many deaths from severe radiation sickness among their colleagues. The repeating pattern of trauma lies in the perception of limits in understanding the events when the witnesses stress that at the beginning the accident was invisible for most of the affected victims and their families. They keep coming back to how shocked they were when they found out what kind of consequences to expect and what could happen to their bodies. As humans turned into highly poisonous nuclear waste, their status changed from private to 'public property'

(Alexievich 2016: 19). Microscopic pieces of radioactive material, the 'hot particles' that killed many liquidators, 'will be capable of killing again' because they are 'immortal' (152).

Through these individual testimonies, Alexievich reconstructs a collective trauma. The affective language that treats Chernobyl as a unique experience shapes cultural memory of it for people who were affected but whose voices were suppressed by communist governments. Alexievich is convinced that Chernobyl is a 'catastrophe of time' (24) and, for her, radionuclides are not neutral ingredients of material reality but horribly eternal in comparison with human lives, especially as they are also invisible to the naked eye. In post-Chernobyl Eastern European memory, nuclear physics is demonized. 'We had touched on the unknown' – says Alexievich – the Chernobyl disaster is 'an undeciphered sign' because 'something had cracked open ... in the space of one night we shifted to another place in history. We took a leap into a new reality, and that reality proved beyond not only our knowledge but also our imagination' (2016: 25). Chernobyl seems to exist more than can be expressed and 'is everywhere, all around us' (225). It persistently clings to the region's memory, stuck to it like Morton's viscous hyperobject.

One of Alexievich's interlocutors says: 'What I'm telling you, it's not coming out right ... The words are all wrong ... I'm living in a real and unreal world at the same time' (2016: 21, 23). The traumatic testimony cannot be spoken – it collapses. Therefore, *Chernobyl Prayer* contains many ellipses, exclamation points and parenthetical notes that indicate when someone is silent or crying. Even the punctuation marks testify to the objectivity sneaking out of the Chernobyl experience. They fill the monologues, instead of knowledge about the accident, as though the reporter wanted to include all the witnesses' emotions, pauses, screams of hurt and sighs of helplessness. The first and the last women's testimonies underscore that even love – the very powerful love of young people who were full of the most sincere desire – had no power over the borderless Chernobyl disaster, since it is a kind of 'dead-end'. Their testimonies are the most emotional because their language falls apart: 'can we talk about this? Put it into words? Some things are secret ... To this day, I don't really understand what this was' (287).

Among the different kind of responses to Chernobyl that Alexievich assembled in her polyphonic reportage, women's voices, including that of the author herself, provide the most intimate narratives. The 'prayer' of the book's title suggests a missing dimension of this traditionally Orthodox Christian community where religion was radically suppressed during the Soviet period. Overall, the muted voices of religious people recall the lack of reporting and censorship about Chernobyl's consequences. Alexievich, therefore, almost ten years after the tragedy, can publish witnesses' monologues as direct reports and advocate the version of events in their post-Chernobyl testimony as a form of moral compensation. She carefully listens to her interlocutors, knowing that the communist authorities ignored Chernobyl victims and did not listen to their version of events. Zink explains the title and structure of the book in a similar way: 'The title of the book *Chernobyl'skaia molitva* emphasizes the phenomenological dimension, for the prayer (*molitva*) is addressed to an absent, higher authority, not to real readers. The monological aspect of the text and the loneliness of the narrators come especially to the fore' (2012: 103). She adds:

> The nuclear accident and its consequences seem to have rendered the limitations of language all the more apparent. Yet precisely the indication of these shortcomings, along with the pauses, ellipses and omissions to which Liudmila Ignantenko [the first female speaker] resorts, are among the special qualities of *Chernobyl'skaia molitva*. The interview partners' silences, their loneliness and their pauses testify to a fundamental rift between things, for they are a sign of nothingness.
>
> 104

This hopeless nothingness deeply affected the post-Chernobyl community, but their voices, collected as a form of 'prayer', still can bring some hope, according to Alexievich – we pray when we believe, even in desolate situations, that there is some consolation beyond human power. However, Alexievich's reportage includes all God's creation and encompasses how Chernobyl disturbed the world of people as well as other beings.

She fills a gap in the history and cultural memory of the Chernobyl catastrophe by paying special attention to what happened to animals, trees and the environment. She says that she speaks of 'the world of Chernobyl' (2016: 25). She does not omit the shooting of all domesticated animals that were left after the evacuation. She lets her interlocutors repeat their descriptions of scenes of non-human carnage – or, in the official idiom, of 'deactivating' or 'cleaning-up' actions. Both human and non-human bodies were categorized as radioactive waste, which taps into post-nuclear apocalyptic discourse. The situation of domesticated animals was especially terrifying: 'Once the villages were evacuated, units of armed soldiers and hunters came in and shot the animals. The horses could not understand what was happening. They were in no way to blame – neither the beasts nor the birds, yet they died silently, which was even worse' (30). She herself remembers when she witnessed such a shooting and heard 'the helpless cries of the animals. They were shrieking in all their different languages' (31). One of the hunters recounts how the dogs, waiting for their owners, were at first excited to see them, 'came running to a human voice', but had to be shot – he rationalizes – as they were 'walking ashes' (110). During the evacuation, the dogs ran after them, trying to get on the buses (107). Eventually, 'they'd stopped trusting humans' (108), 'gone wild' and ran away from people (77, 140). Another person says that she still hears the soldiers killing dogs and their howling (48). Interlocutors recollect how, at the beginning of Chernobyl's aftermath, dogs and cats stayed in yards, 'keeping watch over the empty houses' (88), waiting for their humans and trying to survive by eating anything, even their offspring (41). In Pripyat, evacuated people were informed that they would come back in a few days; a child mentions his pet: 'We left my hamster at home when we locked everything up. He was a little white hamster. We gave him enough food for two days. But we never went back' (273). The breakdown of trust and familial relations with non-humans seems to symbolize a wider, more profound breach that the Chernobyl event caused.

Unlike people, nature could not be resettled. The regional environment went through aggressive decontamination. The clean-up workers saw themselves as destroying the earth in some primary, ontological sense: 'We buried earth in the earth. Along with the beetles, spiders and maggots, that whole separate nation. We buried a world' (104). They removed the contaminated topsoil, carpets of grass with all the creatures living there (196); they pulled up trees and buried them in plastic bags, too (101). The world had changed forever, prompting real ecological trauma: 'as if it wasn't earth; like we weren't on earth' (98). The children of Chernobyl 'began imagining a world without animals and birds' (155), drawing trees with roots growing in the air, with red or yellow water in the rivers, with weeping figures (216). It was 'the war to end all wars', but 'there's nowhere to hide. Not on land, in water or in the skies' (55). Alexievich uncovers testimony about the scale of Chernobyl's disturbance of human–environment relationships and shows how such accounts are an inseparable part of post-Chernobyl cultural memory.

As many testimonies about the exclusion zone indicate, nature has come back there and even flourishes because of human displacement. But for some witnesses, the Chernobyl catastrophe represents an ecocide, not because nature cannot revive there, but in the sense that Chernobyl's environment has been changed from one connected to the human world into an inhuman one, as if it had been denaturalized. In the cultural memory of Chernobyl people, something uncanny was left after the disaster, and it still radiates there and reminds them about the event that turned the world of humans, animals and surrounding nature upside down. The affective and devastating testimonies that Alexievich collected show that pure, realistic reportage based on interviews is enough to recognize in Chernobyl a symbol of human and environmental tragedy in Eastern European cultural memory. However, witnesses' perceptions have to be understood as directly connected with their long period of social alienation living in the Soviet Union's phantom political reality. One of the peasants interviewed said radiation 'frightens people and animals alike. And birds too. And even the trees are scared, but they can't talk. They can't tell you' (Alexievich 2016: 62). Except, she quickly adds, for the potato beetles that propaganda claimed the Americans had sent to destroy the crops, because they are used to Chernobyl's poison (62).

* * *

Nuclear risk narratives that feature the post-Chernobyl cultural memory of Eastern European testimonies serve as reports of how exceptional and unimaginable the catastrophe was, and how it traumatized people and their relationship with the environment. The hyperobjective and affective character of witnesses' narratives was shaped not only by the scale of danger, but also by the lack of political and social security. Chernobyl left a trail of fear, fallout in people's memories like that of the radioactive cloud unpredictably moving for thousands of kilometres like a gigantic kidney over Europe and the world in late April and early May 1986. It left traces in literature by contaminating language. These contaminated elements of language were not hypothetical anymore but started to materialize. These nuclear fears can be detected in other narratives that do not directly refer to Chernobyl but to a certain model of catastrophe, as in the Soviet literature of nuclear risk (i.e. Andrei Voznesensy's *Oza*; Alexander Prokofiev's *Radiation Sickness*). However, it was Chernobyl that created an exceptional, monumental and intimate imaginary of environmental risks within Soviet culture and among the people. Environmental history shows that Chernobyl woke people up very late from their utopian dream – their anthropo-dream. Chernobyl has become a sign of Eastern European risk culture, but it has also become a memorial in Eastern European cultural memory. From an ecocentric perspective, though, it is hard to say how much of a disaster it actually was.

Part Five

Disturbed Landscapes[8]

1

Non-sites of Memory and the Violation of Nature

> By a steep, unfaithful path
> came the exhausted people,
> they carried a landscape on their shoulders,
> the one that lived there with them.
> They brought also with them,
> not knowing about it,
> willows abundant with pears
> standing now behind the fence.
>
> Aleksander Baumgardten, *New steps* (quoted in Kolbuszewski 1985; my translation), 188

Natural landscape features were used by the Soviets to mark occupied territory in Eastern Europe. The Molotov–Ribbentrop Pact with the Nazis, which the Soviets signed secretly in August 1939, laid out borders along the main Polish rivers (the Narew, Vistula and San). As the war ended, around half of the eastern border between Poland and the USSR was kept at the Bug River, while the western border, between Poland and East Germany, was moved to the Odra River. These arbitrary decisions were not anchored in any major historical precedents, since political borders in Eastern Europe had changed often throughout the twentieth century. However, due to Stalin's pressure, during the conference in Tehran (1943), it was decided that Poland would lose its previous eastern territories, including parts of contemporary Lithuania, Belarus and Ukraine, and acquire new territories taken from defeated Nazi Germany. Post-war Poland and its altered borders affected the massive migration of people who were relocated from the East to the West and, what is more, whose native landscapes had significantly changed. They had to adapt quickly to a new political situation and new surroundings but, as in Baumgardten's poem, their attachment to their domesticated, regional 'small motherlands' resulted in a tendency to romanticize the lost, beloved landscape that they wanted to take with them.

Those relocated from rural areas, in particular, began adapting to urban areas by cultivating potatoes or keeping pigs and hens in post-war cities full of rubble and ruins, such as East Berlin (Stoetzer 2018: 301), Warszawa and Bialystok (Szpak 2005).

Landscapes in Eastern Europe were not just affected by repatriation and the strengthened Soviet regime which gained new subordinate nations and their lands after 1945. They were changed most by the atrocities that took place before and during the war; such landscapes lost their pastoral component by becoming the 'non-sites' of cultural memory.

In this fifth and the last part of the book, I attend especially to Eastern European landscapes that are either former military sites or places where major atrocities occurred in natural environments. They have become directly connected with human trauma because perpetrators used the physical features of these landscapes to hide their crimes. I call them 'disturbed landscapes' because both their environmental and cultural status has been critically distorted, first by Soviet and Nazi colonization, then by Eastern European nationalistic politics, and finally by the anthropocentric discourse of cultural memory. In such references to past violence, as recent memory studies scholarship and commemorative practices, nature's romanticized yet active function in constructing the pre-war identity of many Eastern European regions, as well as its ecological presence, is neglected.

To mark how nature obscures memory, Claude Lanzmann coined the term 'non-sites of memory' (1990). The term 'non-sites of memory' underscores the anthropocentric and ethical oppositions between what is human and what is not, between memory and oblivion, between everything that is identified as ours and those sites preceded by the prefix 'non-', which are thereby displaced and thrown out of culture. I want to challenge this predominant vision of environments of history as passive – as non-sites of memory – by first analysing memory studies approaches here in order to propose, in the next chapter, a counter-perspective that rethinks natural sites as places for mourning and remembering with environments.

Eastern European environments are traumatically tangled with memories, histories, ethnicities and ecologies. Human memory, though, sometimes neglects or demonizes the agency of the physical environment in sites of trauma. In order to reveal this kind of anthropocentric approach in commemorative practices and memory studies theory, I turn to places where a cultural representation of the landscape is built into a history of violence. The Volhynia district, Bykivnia, the Katyn Forest, Podlasye and the rural town of Jedwabne, East Galicia,

Kuropaty, the forests around Auschwitz-Birkenau and many other Soviet-era crime scenes demonstrate how essential the landscape is to narrating the history of the region both during the Second World War and its aftermath of mass relocation and ethnic cleansing. However, to what extent does nature witness atrocity? I will examine how the environment as witness is theorized, voiced and muted in memory studies.

In the discourse of memory studies, which has developed extensively in relation to Eastern European mass crimes and genocide committed by Hitler's and Stalin's regimes, a specific cultural construction of nature frequently appears as something that disturbs or even violates commemorative practices and histories of trauma. Within this Romantic frame of understanding nature as cultural landscape, human selves project anthropocentric ways of seeing and position themselves regardless of non-anthropocentric components of nature in sites of memory. In contrast, ecocritical analysis can redirect our attention to understandings and representations of nature and its materiality in places of memory – to their topography, landscape, environment, wildlife, or even organic and inorganic compounds. Especially when gathering nature's beings and their human counterparts in cultural memory results in a promising burst of new terminology and ideas.

Undoubtedly, there are many theoretical efforts that attempt to synthesize traumatic memories with the physical sites of genocidal massacres, mass executions and torture in Eastern Europe. In the absence of above-ground monuments, or in sites politically disrupted by Sovietization, memory must return, either to reconstruct a non-Soviet nation after the collapse of the communist giant, or to give an account of the affective 'aura' (Massumi 2002: 175) of these sites. Physical environments affected by atrocities are construed by human memory into different types of cultural landscapes that have absorbed trauma. Therefore, such landscapes abundantly recreate and proliferate memories. Indicative concepts include the 'memoryscape' where different commemorative practices 'come into contact, are contested by, and contest other forms of remembrance', embodying 'memory at work'. Here 'what memory does', its active role (Muzaini and Yeoh 2016: 9–10), is more important than the place itself. Therefore, 'memoryscapes become focal points around which we build our collective identities in the present' (Kapralski 2015: 150). A similar constructive role characterizes 'traumascapes', which designate 'much more than physical settings of tragedies: they emerge as spaces where events are experienced and re-experienced across time' (Tumarkin 2005: 12). A traumascape is a 'landscape of violence' where 'the memory of violence is inscribed onto space in all kinds of settings', including 'the natural environment' (Schramm 2011: 5), which may even be 'contaminated' with violence (Pollack 2014). All these approaches assume that landscape is reproduced in memory when it represents the topography of suffering and genocide; and all of them can also be connected with the concept of 'affective landscapes' (Stoetzer 2014). However, human affect anthropomorphizes material nature in construing cultural landscapes affected by history – thus they lose their physical, environmental character. Such sites have been disturbed by human violence, but they are also disturbed when they come to symbolize human cruelty and are anthropocentrically constructed as disrupting commemoration. They have become non-sites – places we would like to erase from the map.

In memory studies discourse, geospatial concepts and perspectives multiply the meaning of the term 'site of memory', which is rooted in the original notion introduced by Pierre Nora and later used in such monumental research projects as *Modi Memorandi* by the Berlin Centre for Historical Research of the Polish Academy of Science. The 'site of memory', according to Nora, can be any place that is located *in* memory rather than in the world. It is a vague approach, where 'the site' and 'memory' function as containers of various cultural phenomena and signify almost everything, but they do not exclude combination with environmental memory. Nora's theory serves as a model that always has to be exemplified in a separate narrative, and it happens that the *lieu de mémoire* is inscribed in a specific understanding of nature because of its geospatial character.

When the site of memory designates a place where a mass crime occurred somewhere in the wilds of nature or the surrounding environment, not only is such setting problematic since it is remembered as a place where atrocities were hidden, but so is the question of how to remember it and what kind of memorial it represents. In the poem *Babiy Yar*, written by the Russian poet Yevgeny Yevtushenko in 1961, the natural landscape is so overburdened with the events that happened in this place that it becomes a mere reference point for memory projection:

Over Babiy Yar
there are no memorials.
The steep hillside like a rough inscription.
I am frightened.
Today I am as old as the Jewish race.
I seem to myself a Jew at this moment.
...
Over Babiy Yar

rustle of the wild grass.
The trees look threatening, look like judges.
And everything is one silent cry.
...

1962 [1961]: 82–3

The poet refers to a specific place of extremely disturbing murder – a ravine located in a park area on the outskirts of Kiev, Ukraine – where tens of thousands of Jews, including, for the first time, women and children, were relentlessly executed for two days in 1941 (Snyder 2010: 200–2). There, Yevtushenko did not find any signs of human memory – only the landscape itself served as an insufficient 'rough inscription'. His fear seems to relate to a feeling that the place has been unfairly abandoned and not commemorated, since it lacks 'memorials'. In fact, the first monument, dedicated to all the Soviet citizens who died there, was erected in 1976, while a memorial to Jewish victims with the statue of a great figure of a chanukiah (or menorah) was erected in 1991. There are also plans to commemorate the Jewish massacre by building the Babyn Yar Holocaust Memorial Center in 2021 (Rozovsky 2020). But what Yevtushenko saw in the early 1960s was just an overgrown gorge shaded by trees.

The whole poem's repetitive pattern of being there, 'over' there, 'over Babiy Yar' where 'the rustle of the wild grass' is heard, makes this place sound apparently rather calm and peaceful, but because it is 'there' where the massacre took place, listening to the rustle is disturbed by memory of what happened here. This is a frightening affective landscape – a memoryscape or traumascape, a landscape of violence that is contaminated with history. In Babiy Yar (see Figure 10) the trees are not just trees, they do not stand for themselves: they are the 'threatening' figures of wounded human memory that remind us of violence that took place right here. The poet compares them to 'judges', but what and who do they judge? If the poet can feel like a Jew there, like every Jew who died there, perhaps in that moment the trees represent human evil, the side of executioners. But these 'judges' might also underscore the need to express condemnation and to protest both Soviet silence about the crime and pre-war anti-Semitism – especially because, at the end, the poet recalls pogroms and adds that 'no Jewish blood runs among my blood / but I am as bitterly and hardly hated / by every anti-semite / as if I were a Jew. By this / I am a Russian' (83–4). The natural surroundings of Babiy Yar and human memory merge in a strange but interconnected way when 'everything is one silent cry'. The poet is trapped by a terrifying vision where the discomfort of visiting a non-memorialized place of human evil is intensified by his anthropomorphic projection of guilt or judgement onto nature. The 'wild'

Figure 10 Babiy Yar. Photo by Sam Litvin, Foter.com. CC BY.

grass might facilitate the process of forgetting, while the 'threatening' trees prompt remembering.

'Babiy Yar' exemplifies the binary character of cultural and natural heritage. The distinction between them can result in interfering with nature–cultural bonds in places of memory like Babiy Yar. In the late 1950s, the Soviet authorities decided to build a dam and pump water into the ravine to create a lake that covered the hallowed site. As John Klier noted, 'The result was an ecological disaster. On 13 March 1961, the dam burst and obliterated the surrounding area of Kurenevka. Over 100 people were buried alive' (2004: 290). The Soviets ignored both the cultural and environmental aspects of this place of memory. However, Yevtushenko probably wrote his poem about the place before the dam broke.

Such disturbed landscapes provoke questions about how human memory animates an opposition between nature and culture, which is articulated through various other contrasting categories such as human versus non-human, alive versus dead, rational versus irrational, or wild versus ordered. The environmental layer of commemoration remains undetected in the majority of memorializing texts and studies, which are anchored in a fear of amnesia or in the traumatic tension between remembering and forgetting. In memory studies, nature is often purported to disturb commemoration practices or is treated as a passive, void environment, a mere backdrop for the real events – as a non-site for memorialization. By further analysing such discourse, I ask how ecocritical representation of disturbed landscapes might counter their anthropocentric cultural construction. What is more, many of these places geographically remain the same, but their material nature or environment also changes over time and due to human interventions. However, Eastern European memory studies and environmental historians do not sufficiently address the limits of human physical disturbance, while in the poem, a natural site intertwined with memory, such as Babiy Yar, can be meaningfully silent or can provocatively speak – but for sure, it should not be destroyed.

In the Central and Eastern European region with its rural landscape, thick backwoods and forests, unregulated rivers and wild, solitary meadows are widely inscribed with the traumatic historical events of the past century. While cultural memory transforms these landscapes, their non-inscriptive and environmental status is often marginalized. History and recollections provide a lot of contrasting examples of how, to some extent, the native landscape has been tamed and domesticated for shelter, but it also hides the cruellest Nazi and Soviet crimes, not all of which will be discovered. The Institute of National Remembrance in Poland, which is funded by the government, represents the official narrative of the state's memory politics and vernacular commemoration model. The institution's name is an example of the controversial reduction of memorialization to its 'national' character, while 'remembrance' leaves no space to forget and not commemorate, which is exemplified by a long-term project to investigate former sites of atrocities against the nation. Similar institutionalized memory politics aimed at reconstructing the nation in the wake of Soviet colonization also play out in other Eastern European countries, such as Lithuania. Such memory politics select certain elements of recent history to undergird a monolithic, nationalistic vision, despite the region's multi-ethnic and multicultural past (Moore 2019: 261, 268, 273). Official post-Soviet narratives of remembrance are founded on 'the discourse of martyrdom' and manifest themselves 'in the mass rites of reburial or politics of dead bodies' (258).

The Institute of National Remembrance has been actively searching for the remains of victims of Nazi crimes and of persecution by Stalinist and post-war communists, not only in Poland, but also in Lithuania, Belarus, Germany, the Czech Republic and Ukraine. Its reports give an account of exhumations directly conducted in places of execution, near cemeteries and in political prisons located in forests, clearings or ravines. In the Institute's bulletin and on a specially dedicated page on its website, we find descriptions of many complicated ways of dealing with trees that grow out of and into human remains (IPN 2016). However, none of the historians of the Institute of National Remembrance reflect on the ecological interference involved in unearthing human remains, or consider restoring them to dignity by leaving intact those places of burial that have hybridized with nature. Nature – and mourning with nature – is completely instrumentalized here. However, nature is adored and appreciated as a companion to the soldier's fate in the songs and poems of guerrillas and anti-Soviet so-called 'last foresters' or 'cursed soldiers' – 'cursed' due to the silence about their oppositional activity imposed during the Soviet period. To mention just a few of many famous examples: the boys march and the trees salute in *Szara piechota* ('Grey Infantry'), written by a colonel – Leon Łuskino (1918); over the soldiers' graves, a row of trees hum in Stanisław Bajcer's 'Patriotic Song of the National Military Forces' (1990 [1944]: 135); under a yew tree soldiers dream their last dreams before their sacrificial death, in which the tree substitutes for a cross, in the anonymous song 'Partisans' (1952: 10–11); and 'the carpets of forests and meadows' seen from a prison window symbolize freedom in Antoni Makarski's 'About the Soldier' (probably written after 1948, when Makarski himself was imprisoned by Soviet authorities, Narodowe Siły Zbrojne 2010 [National Military Forces]). Then there is this famous song from the First World War, 'On the Ashes', written by Edward Słoński and sung by anti-communist fighters, where flowers and herbs, not weeds, will grow and co-memorialize the soldiers' graves:

> Those fields the enemy was trampling
> and ploughing with fire,
> God will sow
> with cornflowers and camomile in the spring.
>
> The fragrant herbs will overgrow
> around the soldiers' graves,
> sharp scythes will call
> in our fields and meadows.
>
> And cornflowers, and camomile
> crowning each grave,
> to the soldiers who died
> they will bow to their feet.

They will bow very low
circling these sad tombs
on this torn ground,
on these beaten meadows.

Słoński 1990 [1914]: 163–4; my translation

These well-known domesticated Eastern European plants, with their divine vitality, play a different role than the trees that are 'judges' in Yevtushenko's *Babiy Yar*. In Słoński's song, nature participates in mourning and commemorating the numerous anonymous graves that cover Eastern European ground. The flowers stand in for erected monuments and memorialize those below the ground, turning the battlefield into a colourful site of memory (if seen in the spring or summertime).

In these Polish patriotic songs, anthropomorphized and appreciated nature mitigates the heaviness of the traumatic experiences soldiers had when they hid in forests or other natural areas during the wars and in the communist period after 1945. Despite the popularity of these lyrics among consecutive generations of Polish soldiers and anti-Soviet fighters, the state committed itself not only to relentlessly searching for sites of Polish martyrdom but also to 'exhuming and relocating the remains of Poles who died or were killed outside the country to the cemeteries in Poland' (Institute of National Remembrance website). Natural areas do not suffice for commemorating former military sites. The possibility of natural memorials is ignored in the official discourse on commemoration.

The problem of subjugating nature is theoretically and poetically reflected in anthropocentric memory discourse when it uses Eastern Europe's landscapes to represent either traumatized memory at work or how uncommemorated locations of atrocities are forgotten. Nature's inhuman vitality signifies either amnesia – as Martin Pollack observes in his *Contaminated Landscapes* (2014) when discussing sites of mass crimes – or that nature's vitality baffles the attempts to commemorate victims of mass murders or even genocide (Lanzmann 1990; Snyder 2010). Natural landscapes cover the past from view, they represent 'places of absent memory' (Baer 2002: 66). And 'it is not the driving force of symbols – of signs posted, or of tombstones – nor the language of ruins. Here nature covers over, transforms, and does not allow the visitor to view the past' (Sendyka 2015b: 14). Thereby, as I interpret it, nature can attract hatred. Whereas affective human memory, dominated by trauma, seems to operate heedlessly or antagonistically towards these sites by neglecting their environmental history, and to ignore that, for example, clearings made by Nazis in the forest near

Sobibór do communicate how this landscape was destroyed. The relation between human and environmental violence here is not, though, recognized by cultural memory.

How does nature complicate the discourse on places of memory and genocide, or killing sites in general? What does this mean, especially in light of the centrality of Pierre Nora's theory of 'sites of memory' – which he understood as 'places about memory' or 'places where memory works' – for memory studies? After all, the construct of the place of memory – as Maria Kobielska pertinently noted in her article on *Les Lieux de Mémoire* – focuses not on what we remember but on the active process of reworking the past in memory, not on the given, passive material of tradition but on constantly expanding memory (2013: 195). We do not speak about the places that contain a memory of an event but about the activity of human minds, about attracting their attention and engaging them in taking care of such places (196). Can we also take care of physical nature there, as in the case of forests, and expand the meaning of place of memory beyond the anthropocentric construction?

Theorists of cultural memory studies, such as Astrid Errl, Michael Rothberg, Jan Assmann and Aleida Assmann, recognize different memory media, including landscapes. In other words, in concepts such as 'memory landscapes', landscapes serve as semiotic media for cultural memory (Assmann 2011: 44). While some meanings of landscape – a contingent term for a site of memory located in natural environment – consider how the ecosystem would function independently of the meanings imposed on it (Kühne 2017: 30), environmental memory is not part of cultural memory in memory studies discourse. Barbara Frydryczak, a leading scholar of cultural landscapes in Poland, states that nature – in the case of cultural landscapes – is just material that is formed by culture. She includes elements of natural landscapes in her definition but they do not form an essential part of human–environment transformation: 'cultural landscape results from human work, activity, and historical period. The traces of these actions are still readable. It is possible to recognize them in the landscape topography (roads, fields, trees, alleys) by uncovering the following layers of meaning (landscape archeology, stories, legends), as well as in the artefacts (monuments, memorials, places of memory), which implies going beyond nature in the direction of human historical testimony' (2014: 198). On the other hand, when nature is not taken as a mere material of human activity, understanding a cultural landscape as a construct or a text, which is shared by cultural geographers (on the concept of reading the landscape, even through 'nature', see Traba 2017: 14), does not have to exclude the role of physical environment in memorializing trauma. But when the affective, unsettled and inhuman concept of landscape seems to dominate in recent Eastern European memory studies, thanks to its capacity to evoke human trauma, it means essentially cultural projection with little or no environmental frame of reference. Such anthropogenic character of the cultural landscape belongs to a certain narrative paradigm in which culture and nature have already been separated. Nature is denaturalized, disturbed not only by history but also by the anthropocentric, affective memory of Eastern European discourse, or, in other words, the non-human is humanized.

In this type of memory discourse, the natural surroundings – a defining component of places of genocidal massacres, executions and torture – exist in opposition to the prioritized cultural need to remember. Nature functions here as an obstacle to potential commemoration – a disturbant or malicious, anthropomorphized agent of forgetting. After all, these wild, inhuman sites of memory have not been memorialized with museums or monuments. As a result, the natural site of memory is anthropomorphized as 'a guilty landscape' (van Alphen 2000, in Sendyka 2015b: 19); a revenge of 'abundant vegetation' (Sendyka 2015b: 17); a 'shamelessly alive' environment (Sendyka 2015b: 28); or demarcated with death to cover the whole Eastern European landscape (Baer 2002: 72).

Martin Pollack exemplifies how this memory discourse blames nature for forgetting. In *Kontaminierte Landschaften* ('Contaminated Landscapes') he observes that many landscapes of mass crimes in Eastern Europe lost their 'would-be innocence' even before the birth of National Socialism in Germany (2014: 14). The landscape, 'contaminated' with human violence, literally becomes co-responsible for human forgetting of abandoned mass graves (Pollack 2014: 20) because it is a hidden place (Pollack 2014: 27), a place ready to be used by the perpetrator, a place that will always surprise by looking like the land of one's childhood (Pollack 2014: 37). According to Pollack, we must not stop asking, 'Does *this* landscape have something to hide from us? Is it really so innocent and idyllic as it seems? What will we find when we start digging here?' (2014: 54). What is more, landscape is potentially always condemned when it is located somewhere in Central and Eastern Europe, as are other elements of the natural environment. Pollack gives even the example of water, which can also 'be a co-perpetrator like the landscape' because, as he says, it is 'silent as the grave' (2014: 60). In fact, he essentializes this region's environments and incorporates all their physicality into a theory of contaminated landscapes in Central and Eastern Europe. It is striking that a term that has ecological connotations does not lead him to any non-anthropomorphized reflection on sites of nature in the history of violence.

Nature incorporates violence (Smykowski 2017: 65, 74), despite being depicted at times as a victim and a shelter (Czapliński 2017: 11; Smykowski 2017: 75–6). The essentially affective character of these sites, because of the Nazi and Soviet locations for atrocities, stigmatizes Eastern European landscapes and can lead to material nature's exclusion from cultural memory. Following Lanzmann's term 'non-sites of memory', the theory how nature obscures memory and commemorative practices has since been further developed by Roma Sendyka. Under the 'affective aura' of these non-sites (Sendyka 2015b: 16), it is not memory that works but poetic language: nature and desolated landscapes 'contaminate' and 'overgrow' (Sendyka 2014: 88), eradicate and hollow out (Szczepan 2014: 124) the memory that needs to be purified. While affective language merges with experiences of the haunted space of the non-site of memory, nature in itself is made invisible. Indeed, Sendyka comes to the conclusion that, in situations where there are no 'ruins', 'monuments' or any otherwise readable 'landscape', nature cannot memorialize (2014: 101), even though she knows that these landscapes do contain physical traces of atrocities.

Similarly, Aleksandra Szczepan (2014) reconstructs nature's neglect as intrinsic to the category of the post-memorial landscape, which is still more affectively loaded. She bases her analyses on works of art and literature that present a displaced narrative of the 'sinister' environmental settings of the Holocaust. Again, the non-site of (post-)memory obscures evidence of tragic history (Szczepan 2014: 111), while nature is accused of being a part of *this* crime of erasure (Szczepan 2014: 122). The landscape is seen as 'a screen for traumatic recollections' (Szczepan 2014: 110), but there is nothing supportive in this vision – it is primarily a hostile image of nature, characterized by a 'misleading aura of normality' (Baer's concept), where the disturbing agency of nature is contrasted with the 'comforting' space of preserved camps (Szczepan 2014: 111). These post-memorial landscapes of nature lose their geographical and environmental identity. They become empty, alienated places (Szczepan 2014: 120), where nature plays a strange, isolating role and is reduced to a mere anthropomorhized sign.

'Placing the blame' on the anthropomorphized landscape for preventing us from memorializing loss at sites of mass crimes – as Pollack does – is part of constructed stigma covering landscapes in Eastern European cultural memory. Such memory constructions can disturb *our* human relationship with nature, and exclude ecology and environmental protection from managing those sites of memory. Natural areas are also subjected to historical transformations and destruction while serving as sites for mass killings and preparations for genocide (Domańska 2017). However, nature's vital resilience, the way it regenerates even in the places of the worst atrocities, the way it grows back over them, portrays nature as cruel and alien to the human need to commemorate. In literature, however, this vision of nature as interrupting commemoration in memory studies discourse seems to be more complex and needs to be reconsidered.

An example of commemoration reflected in nature is Czesław Miłosz's poem *Odbicia* ('Imprints' or 'Reflections'). After human losses, greenery revives in the 'field of cornflowers' which 'blossoms after the fire' (1987 [1942]: 202). Miłosz juxtaposes the tragic history of a city (Warsaw) and its residents during the Nazi and Soviet occupation with nature, and challenges 'a literature then accustomed to a noble, martyrological tone' (Fiut 1990: 42). In the non-human world, death is a natural phenomenon ('a trampled ant', 'a dead field mouse'), but because of the context of traumatic human history, this correlation dismays. Nature may be perceived as radically alien when it regenerates despite and after human tragedy, but it also calms and carries some hope when the poet echoes at the end: 'This is the image that is reflected in water' (202).

Another example comes from Tomasz Różycki's collection *Scorched Maps*, which is comprised of travel poems to sites of atrocities in Ukraine. The title poem uncovers a multilayered complexity of vegetation and memory:

I took a trip to Ukraine. It was June.
I waded in the fields, all full of dust
and pollen in the air. I searched, but those
I loved had disappeared below the ground,

deeper than decades of ants. I asked
about them everywhere, but grass and leaves
have been growing, bees swarming. So I lay down,
face to the ground, and said this incantation –

you can come out, it's over. And the ground,
and moles and earthworms in it, shifted, shook,
kingdoms of ants came crawling, bees began
to fly from everywhere. I said come out,

I spoke directly to the ground and felt
the field grow vast and wild around my head.

2013, translated by Mira Rosenthal

The abundant vegetation, the excess of life that suddenly appears on the borderland's meadow, full of the corpses of loved ones, may indicate the necessity of getting down close to such below-ground sites of memory. Simultaneously, nature is depicted as strange, 'wild' and distant from traumatic human history and the need to commemorate ('grass and leaves have been growing, bees swarming'). However, the way in which the separate stanzas structure the poem, as if pausing the human traumatic memory or opening it up, inevitably links human with non-human to show a different dimension of life – one that is above, with, next to and below the remains that are to be memorialized. Nature fills the gaps in memory with strange sensual vitality and resilience, and this is combined with the speaker's embodied experience of laying down and touching the ground. The paradoxical feeling of nature's material presence in lively sounds and movement contrasts with the site of death. It is surprising that a site of genocide can be so full of life. An experience like this can be painful and isolating in the absence of a cemetery or memorial; but though the poet shows how nature responds, he does not reveal whether it says anything.

Nature is often used to refer to genocides in Central and Eastern Europe, as when its fauna and flora function as symbols of human trauma [Sendyka 2015: 83–4]), e.g. the hares in Lanzmann's *Shoah*, the roes in Bałka's *Winterreise*. However, in contrast to Miłosz's and Różycki's poems, nature is also *misused* in the anthropomorphic stylistics through which theoreticians such as Pollack lay blame on it, inadequately projecting it as a co-perpetrator, an untouched observer, or a silent witness to crimes. Sendyka writes that, in the artwork *Der Baum* by Erik van der Weijde, trees are *uneventfully* (peacefully?) *looking* at the family house of Hitler (2015: 84). She concludes – quoting Baudrillard's *Cool Memories* on the indifference of trees to the historical moment of 1990 (2015: 87) – that nature turns out to be not only unaffected but also *not interested* in interrupting and ending the execution of humans (84). These haunting (non-) places of memory are considered radically inhospitable to human identity: 'there is no possibility to inscribe oneself into this space, to domesticate it, make it your own' (Sendyka 2014: 90).

The habit of anthropomorphizing nature is anchored in the developmental process of the child (Tuan 2013: 7) and may later serve as a control mechanism to cope with negative emotions, such as fear. Tuan's concept of 'the landscapes of fear' applies to the 'non-sites of memory' introduced by Lanzmann and described by Sendyka, and to 'the affective landscape' discourse in memory studies, since they are all 'the almost infinite manifestations of the forces for chaos, natural and human. Forces for chaos being omnipresent, human attempts to control them are also omnipresent' (Tuan 2013: 6). However, this does not justify the material, ethical and ecological consequences of the *discursive violation of nature* – a kind of obsession with projecting anthropomorphic features onto nature. The way that cultural memory blames, ignores or anthropomorphizes nature erases the environment as an environment (in its non-human, physical character), which is indicated in the antithetic term 'non-sites', 'non-places' or 'strangely placeless' sites of memory (Baer 2002: 69). Sites of mass crimes abandoned by humans belong to the landscape of forgetting (Sendyka 2015: 94), but if they are not devastated or ruined by humans, they may also serve as sites of memory preserved in the environment.

Nature cannot be erased from cultural memory because it is perceived as haunting the memories and because it is constructed as an obstacle to commemoration due to what happened in Eastern Europe. Perhaps, strengthening the environmental reason to protect nature in the sites of memory, can be an act of opposing the history of Soviet and Nazi occupation, and emancipate nature from these dark anthropocentric memory constructions. Places of memory can be understood differently and do not have to be subjected to the ecological violence of memorializing. Therefore, the next chapter is devoted to landscapes of memory that are not merely passive backgrounds on which historical events are inscribed. Traumatized memory anthropomorphizes the presence of nature in these disturbed sites, while what nature communicates there is the question of the next chapter.

2

Greening Sites of Memory

The construction of sites of atrocities in Eastern Europe as affective landscapes, and their aura, has significantly influenced cultural memory. For example, the Soviets' first photographs of the Auschwitz-Birkenau camp, just after its liberation, with black barracks and barbed wire standing out against snowy white winter, came to represent the whole landscape of Nazi genocide and to figure cold, rational and ruthless evil: 'Just as we both expect and want the perpetrators to look evil, similarly we wish to invest the landscapes of the Holocaust with evil and tragedy,' wrote Andrew Charlesworth (2004: 218). In Mirosław Bałka's *Winterreise*, parts *Bambi I* and *II* – often interpreted as representing places of Holocaust (non-)memory because of their winter scenery – depict Brzezinka's forested landscape in which little roe deer are seen behind barbed wire. Despite varying interpretations of the artist's work, it is clear that the animals are nonetheless entangled in this place; they look trapped and afraid, as the artist admits (Bałka 2004: 29). Should we choose to see it, the history of this site of memory because of the genocidal past evolves into a dark environmental history that continues in the present.

Experiences of landscapes disturbed by violence overturned pastoral and romanticized images of Eastern European nature. However, the environmental aspects of these sites and their meaning for memorializing trauma has a huge potential to depoliticize, denationalize and green memory in this region of conflicted histories. Such possibilities for cultural memory have not yet been discussed from an ecocritical perspective. Including (material) nature in the process of mourning and memorialization can reconnect people with these stigmatized environments disrupted by atrocity and suggest other ways of expressing wounds and trauma. By rereading these landscapes and literary sources and by looking for what Val Plumwood called 'respect for material-ecological traces' in memorial practice (2007: 61), and treating the environment, as Jessica Rapson's suggested, as a 'medium for remembrance' (2014: 161–71), I aim to green these sites of Eastern European cultural memory. Especially in literature and testimonies, I find alternative ways of commemorating that do not rely on built monuments and ordered cemeteries full of artificial flowers and plastic kitsch. From an ecocritical perspective, I am interested how material nature itself intervenes and participates in these sites: what does nature communicate beyond these sites' affective and politicized dimensions? In the cultural memory of Eastern Europe, it is rare to recognize the multidimensional character of sites of violence where human suffering is entangled with environmental actors. But it is worth revisiting the most contentious places of memory because nature can remediate deeply nationalistic strife. One example includes a region where the Second World War triggered a dramatic ethnic conflict between Polish and Ukrainian communities.

The historical rural region of Volhynia – which is now part of *Polesya* in Ukraine – is called the 'Polish heart of darkness' (Huk 2013) because of the genocide perpetrated by the Ukrainian Insurgent Army (UPA), mostly between May and July 1943, on Polish citizens of various ethnicities. The testimony of survivors indicates that they sought shelter in the forest (Motyka 2016: 142). Members of the UPA similarly hid there from Poles who were seeking revenge as well as from Soviet soldiers who were chasing them through the forest. Polish soldiers in the partisan army (those who fought in irregularly organized troops, some came under the Soviet army, others belonged to the anti-Nazi and anti-communist units like the Home Army) and Jewish survivors (Cole 2014) also recall using the forest as a shelter.

In Eastern Europe, victims or perpetrators are part of one's ancestry, and trauma is understood as passed down just like genes. In my own family, there is a deeply hidden horror story of a Polish–Ukrainian misalliance that led to the murder of my great-grandmother and her nine-month-old child during the so-called Volhynia Slaughter. My grandfather managed to escape with his father and his other siblings. They hid in the forest and nearby meadows. Then he became a partisan (at first, he joined the Home Army's resistance), and later a professional soldier (in the communist partisan force of the Soviet-dependent People's Army). He never spoke about what happened, but the story was reconstructed by witnesses, survivors and their children (Horoszkiewicz 2016). What strikes me today is the peaceful, spacious rural landscape where the massacre and burials took place, with some single trees rising over the high gold grass. It is reported that when some attempted to dig the trees up, the Ukrainians who now live there protested, and the burial place was protected. What is more, a Ukrainian living on a farm that used to belong to Poles did not want to cut a lilac tree there, saying that 'that's all that is left of the Poles, let it grow'

(Horoszkiewicz 2016). Despite the history of the Volhynia Slaughter, which is painful for Poles and Ukrainians and has been permeated with the two nations' conflicting and unresolved memories since the collapse of the USSR, landscape and nature entangle here in releasing shared trauma.

In order to counter anthropocentric approaches to commemoration, cultural representations of landscapes need to encompass how nature relates to historical events and how it is co-remembered in Eastern European memory. One such memory site in the forest is the location of the Katyń Massacre, one of the most traumatic crimes committed by the Soviets in Polish history. In Eastern European cultural memory, Katyń Forest has become a major symbol for a whole series of events that took place there and in a few other wooded areas, where multi-ethnic Polish elites – including major officers and highly educated people who were taken from the reserve forces – were executed at the beginning of the Second World War. These forests were located in the USSR near its border with Poland and are named for nearby towns and villages, such as Smolensk, Mednoye, Kharkiv, Bykivnia and Kurapaty. The Soviets had already used some of these places – including Bykivnia, which is now in Ukraine, and Kurapaty in Belarus – to secretly bury victims of regime from various ethnic groups. Since all these massacres were carried out on Stalin's orders, the propagandistic communist media was not allowed to investigate the real perpetrators. For almost fifty years, the history of executions in these forests was only spoken of in the underground or in exile. The same was true of Katyń, although mass burials there were discovered early, by Nazi German troops in 1943.

While Katyń Forest is now portrayed as a specifically Polish national symbol of martyrdom, it is part of a range of similar sites that Stalin's relentless purges left in other nations of the former Soviet Union. What binds these places together is that, in each of them, a specific environment was used to hide crimes. Therefore, they all signify a certain type of violence committed in a natural area – one where all the victims and potential witnesses were dead, except for the executioners.

The question of the witnessing that literature undertakes by speaking for the natural landscape and including it in expressions of trauma has not been studied from an ecocritical perspective in memory discourse on Katyń. Scholars have noted, however, that Katyń represented a lacuna in memorialization because of the lack of testimonies about it. Therefore, all the objects that were found on the sites during exhumations – not only human remains but also scraps of clothing and personal belongings – are treated as relics (Blacker 2012: 114).

The poems I analyse here can be seen as integral parts of producing memory through nature and can serve as indirect testimonies that fill the emptiness, or what Lanzmann, Baer and Sendyka called 'placeless' events in the case of the Holocaust non-sites of memory. Some of them were written by victims' relatives, who could not organize any funerals but included the forest and its environmental surroundings in their mourning. The affective aura sensed at these sites is even substituted with the figure of nature's grief. This is the case with poetic memorials created by daughters of the murdered officers.

In the poem *Impresje Katyńskie* ('Katyń Impressions'), for example, Alicja Patey-Grabowska creates this powerful image of nature mourning 'for a long time'. In the situation of Katyń's victims' relatives not reconciled with this atrocity, feeling angry at how the perpetrators were kept secret, and at how long the crime has been hidden in official memory during communism, the voice of environmental memory sounds loud and rebellious against the censored history. She writes:

> Are there no witnesses?
> And the sky
> penetrating and angry
> The trees were groaning for a long time
> to bawl this crime
>
> Patey-Grabowska 1989: 20; my translation

And Witomiła Wołk-Jezierska, in *Garstka ziemi* ('A Handful of Soil'), shares this intimate experience, how some ordinary pieces of forest can turn into relics. And perhaps they can contain even a tiny organic material of the murdered father – 'an atom of bone'. In the absence of a corpse for a Catholic funeral, this belief that the forest preserved a particle of her beloved father can tranquilize a disturbing trauma of loss that left her and other families in the 'endless waiting' to pay the last respects to the Katyń victims:

> After endless waiting
> I have you right next to me
> in a handful of soil,
> in a pine needle,
> in a blade of grass,
> an atom of bone – you came into being.
> The half-century of waiting blessed
> the patience of pain,
> the endless helplessness,
> the abyss of silence
> – permitted to pray loudly ...
>
> 2013; my translation

Similarly, Halina Młyńczak creates a poem-prayer in memory of her father and his comrades, in which the wind whistling in Katyń birches plays a part in commemorating and releasing the trauma. When the family comes to visit this site of atrocity, not sure if it is the right spot for the murder of their loved one, their experience of this place is like being at a cemetery, where they can peacefully pray with nature:

> In the Katyń Forest
> The birches whistle for you
> Eternal God
> Give rest!
>
> In the Katyń Forest
> The sun rises
> Good God
> Give eternal rest!
>
> 2002; my translation

A former soldier and émigré writer, Feliks Konarski, also addresses the lack of human witnesses to reconstruct the history of Katyń sites of memory. In the poem *Katyń*, he animates nature as part of the memorializing process where trees stand like 'blessed candles', 'birds of sorrow' do not sing or call but 'wail' like weeping women during the last goodbye (especially in Christian rural communities of Eastern Europe, such women were hired by the family of the deceased to pray and weep before the corpse was transported to a church). Environment is the first archive of the crime, the most direct source of knowledge. And even if it is a secret place, nature seems to be anthropomorphized as willing to help to uncover the truth in 'looking for' traces – for the 'signs of former life'. Willing to hide the crime was the Soviet perpetrators' aim. Nature symbolizes life in this wild cemetery, not death. The author is upset that nature cannot tell how the crime happened, but perhaps it is a rhetorical strategy to express his bewilderment:

> Because only the trees above the grave
> Standing like blessed candles
> Could through their leaves' rustle
> Whistle this secret ...
>
> Because only this silent earth,
> Covering captive bodies,

Could confess this cruel truth
Could if she knew how.

...

Today, only the birds of sorrow
Wail miserably in the forest,
As if they remembered
About this Katyń Spring.

As if they were looking
In the forest's underbrush
For traces of captive death,
Signs of former life.

Whether from under the oak leaves
Or pine needles
An officer's epaulette will shine
Or a rusty eagle emblem

1998: 1; my translation

The trees and birds 'singing a requiem' also become witnesses and figures for mourning in Waldemar Kania's *Mord* ('Murder'), where the human and ecological trauma are combined via the metaphor of chopping the trees down. The crime resembles a more-than-human catastrophe that shook the ground, so mighty and shocking it seems that it cannot be expressed within a human construction of the world:

In the forest
trees were witnesses
mutilated
other witnesses were cut down
a man was the executioner

...

birds were singing a requiem
and the earth was trembling
at mass graves

1989: 6; my translation

In the absence of man-made memorials, nature memorializes the site of atrocity, as in Anna Rodak's poem *Las* ('Forest'). Since Katyń woods cannot represent an ordinary forest anymore, the memory of the crime is stored in the environment and detected in 'different' sounds of the forest, commemorated by the 'living monuments' of trees. The trees still grow there despite the blooded ground, but, as if in respect for what had happened, 'in silence'. This way Katyń nature becomes special and precious because it keeps memory alive. In such sites, traumatized memory projects onto nature the ability to actively participate in cultural memory, although by listening intently into nature's own repertoire of expression. Even though the forest cannot speak, it communicates the history of traumatic past and strengthens the connection between traumatized people and physical environment. This history cannot be found 'in any book', but the forest can be read, and the message about 'those murdered, buried, and forgotten' is received:

> Somewhere far away,
> or maybe quite close,
> there is an old forest.
> A forest like many on earth
> full of trees and singing birds.
> And yet, this one is ... different.
> Here the trees rustle differently
> the birds sing differently.
> If they could speak
> they would tell a story
> you won't find in any book
> that no one will tell anymore.
> So, they grow in silence
> on the Katyń soil
> wet from blood
> and shed tears.
> Like living monuments
> of those murdered, buried and forgotten.
>
> 2015; my translation

Katyń Forest and many other similar natural sites of violence represent disturbed landscapes and speak through literature, in its own voice, in cultural memory. Not only the physical forest and its surroundings were disrupted by Stalin's purge and decades of communist censorship, but also from the perspective of Eastern European cultural memory – the pastoral or romanticized tradition of representing nature (I come back to this trope in the next and last chapter of the book), according to how this tradition was differentiated by Lawrence Buell (in Gifford 1999: 4) or Terry Gifford (1999: 2–4). What violence did to those natural sites and forest landscapes, as expressed in Katyń literature, changed the pastoral trope of celebrating ahistorical nature always in contrast to urban or human history, or civilization. However, in Katyń poetry nature is celebrated as a guardian and the responsive side of cultural memory, and perhaps, thanks to this, it creates a new variant of pastoral commemorative literature when it becomes a physical and symbolical part of history. Moreover, in Katyń poems, memory preserved in nature helps to rework trauma, shaping it and facilitating commemoration of the victims. Literature enables environments to speak within this new pastoral tradition, instead of constructing them as affective non-sites of memory that disrupt mourning.

East European forests were a primary setting for atrocities during the war; however, one can also find examples in which bodies were thrown into rivers. The most famous river of the Holocaust is the Soła River, which carried ashes from Auschwitz-Birkenau's crematoria – it features in a scene in the film *Son of Saul* (Nemes and Royer 2015), which was inspired by the *Scrolls of Auschwitz* (documents written by members of the *Sonderkommando*). The Soła also provided a way to escape, as gunshots could not always reach the many people who decided to swim across the river to the forests. Now, the Soła flows into three artificial lakes created by dams. They were constructed because the river kept destroying nearby settlements and had to be tamed, especially after 1958 when it flooded the city of Żywiec. The Soła is considered one of the most unpredictable and dangerous rivers in Poland, and memory of the floods is preserved by residents (Żywiec.info 2014). However, it is hard to find information about the Soła and its role in the Holocaust. In the 1970s, when the Soła was used to build a huge hydroelectric power station – one of the most ambitious hydroengineering projects during the communist period – the expanded reservoir also obscured any evidence of the Soła's participation in the traumatic history of Auschwitz-Birkenau. The only existing project that incorporates the river into Holocaust commemoration is a recent initiative to build a special bridge formed as a ribbon over the river to connect the banks of the city of Oświęcim and its specially designed parks with what remains of the site of the former camp (Oswiecim.pl). However, what happened with the Soła under communism, as well as with other environments that have not become part of cultural memory like Katyń Forest has, prompts an ecocentric argument about how the material nature of memory sites must be protected.

Today, scholars such as Aleksandra Ubertowska (2015) and Przemysław Czapliński (2017) recognize a real parallel between historical and environmental harm in the case of the Holocaust in Poland (Małczyński 2017: 31). But this parallel also applies to other mass crime sites that are closely intertwined with images of the Eastern European landscape. In this case, by searching rural areas named as the killing fields or sites of relocation (e.g. in Belarus and Ukraine), people can find forgotten traces of atrocities in natural landscapes. Nature therefore has an affirmative, not obliterating, role in the act of remembrance.[9] Andrzej Stasiuk, in *The East* (2014), grasped this phenomenon in his own way and named it 'the memory of sites' and the 'memory of landscape' (Stasiuk 2014: 14–15), which could not refer to any landscape but this distinctively Eastern European environmental history formed by 'bloodlands' (Snyder 2010) – dumping sites for corpses and history (Stasiuk 2014: 88–9), a land of human remains (Stasiuk 2014: 112) that have been biologically transformed:

> In the East, on the right side of the Vistula. In Hell's Mouth. On a mound of dumped human remains. In the darkness of the continent. You can't disconnect from these lands: they are like a layer cake of muscle, blood and bone, infused with DNA.
>
> Stasiuk 2016

He does not accuse nature of hiding or 'contaminating' such places as Martin Pollack (2014) does, but perceives it as an integral part of memorializing practice. A disturbing landscape, from an anthropocentric perspective, means a natural landscape that is considered an obstacle to memory, while from an ecocentric perspective, a disturbed landscape is either one contaminated by mass crimes and their traces or one where nature has created a unique material and biological bond with human remains that are or might be located there. Thus, discourses about the pollution of environment and of memory intertwine.

How does nature create memorials through its very fabric? Or how does nature remember? We can imagine ash and bones and human flesh disintegrating into nature and contributing to the ecology. Why do memorials have to be manmade? Is it because that's the only way we understand commemoration? Or rather, does Eastern European history enable us to recognize natural landscapes' potential to produce environmental memory within cultural memory? Does it prompt us to seek non-anthropocentric connections, as in studies of 'the environmental history of the Holocaust' (Cole 2014; Małczyński et al. 2020).

In this context, research conducted by Jacek Małczyński is worth recalling. The scholar often refers to nature as an irreplaceable agent in commemorating the Holocaust of the Jews. He unburdens the discourse about natural sites of memory of its anthropocentric overload. He seeks life where others see only death (compare his analyses of works by Bałka, Hueckel and Sliwiński, and Hansen). He defends nature when others fight it, such as the clearing of so-called 'weeds' by the Museum Auschwitz-Birkenau (Małczyński 2013). Nature in

Holocaust sites may often be the only carrier of memory[10] for future generations, he concludes (Małczyński 2013: 237). This leads to his observation that the memory of genocide is partially – sometimes even materially – deposited in the living tissue of nature. Thus, the site of memory belongs not only to the cultural realm but also to the environment; it is as a hybrid space where new material and narrative bonds are created between natural and cultural worlds that had been in conflict. Similarly, Plumwood reflects on cemeteries as places where ecological communities include both humans and non-humans (2007). Thus, mass graves and genocidal resting places located in nature can serve as a strong argument to protect even the most disturbed landscapes against destructive intervention into their ecosystems.

However, there is one more serious issue that should be discussed in relation to greening the narrative of places of memory. Archaeologists such as Caroline Sturdy (2015) have raised this problem of decomposing human corpses (also discussed by Ewa Domańska in her latest book *Nekros* [2017]). On the one hand, it is well known that a natural landscape's features are an important part of 'the architecture of oppression' (Sturdy 2015: 241). All sites of crimes and genocide constitute 'material witnesses' (Sturdy 2015: 265) – places where the evidence left behind somehow speaks, though it is not always tangible, and sometimes, despite society's expectations, cannot be excavated. That is why we may want to treat these sites of mass crimes as memorials that should not be touched because their material dimension of environment, and how it affects memory and poetic imagination, warrants embodied experience of commemorating the victims, as in Katyń poetry and the memories of relatives who visited those sites.

According to forensic research we know that human remains can be detected in physical environment, and cultural memory reworks this biological presence in nature in its own way. Poeticizing environmental memory can be helpful in preventing those physical sites of memory from the aggressive intervention of exhuming bodies and can suggest other commemorative practices. Despite the Nazis' rationalized projects to colonize, order and modernize Eastern European environments, the wild, unkept landscape remained under the discourse of protection (Blackbourn 2006), and Eastern European nature survived in its stubborn, vital form. This has been neglected by anthropocentric memory discourse; however, nature as an agent opposed to Nazi plans belongs to the history of not only human but also non-human resistance to violence. Despite everything, the trees still grow on Katyń Forest's sites of memory and are perceived as witnesses.

In the introduction to a special issue of the journal *Teksty Drugie* devoted to the environmental history of the Holocaust (2017), Przemysław Czapliński argues that the environment was neglected in crime sites, but in present reflections is considered as 'an entanglement of voices and evidence' (2017: 12). He asks an interesting question: what language should be used to represent these 'multilingual human-nonhuman ontological collectives' (Czapliński 2017: 13) in places of mass murder and genocide? To continue this approach further, we can ask: how should we narrate the environmental history of the Eastern European region in order to more adequately represent its natural lands as indissociable from its traumatic and contested cultural memory? Such books as *Kontaminierte Landschaften* and *Bloodlands* study these events only from the perspective of anthropocentric history. 'More adequately' means, here, from an ecological perspective that accounts for Eastern European environments and their particular landscapes, waters, mountains, forests and meadows, as well as their indelible figurality and materiality for mourning. We can look to cultural texts, like the Polish soldiers' songs and poetry on Katyń, that redirect memory practices away from representing humans as the only victims of war and communism in this part of Europe. Such texts instead prompt us to include nature as a material and spatial participant in these histories. However, the way that we perceive death and funeral practices influences our environmental culture and modes of commemoration. Therefore, the biocentric narrative may be too radical for Eastern European traumas, which are still dependent on an affective and anthropocentric construction of nature in these sites of memory. Nevertheless, let me give an example of an affirmative and celebratory communion with ecological others that provides another way to overcome the trauma of death, especially given that the author, analysed earlier in the book as a critic of collectivization, was a witness and victim of Stalin's purges, sent to gulags in Siberia and other hostile places.

In his poem *Metamorphoses*, Nikolay Zabolotsky shows how 'our superstitions' are based on the individualization and humanization of death. In the process of dying, natural 'forms' are involved. They carry the human body's matter and make it part of some harmonious network, 'a living architecture', as musical instruments create 'a singing organ'. There is nothing frightening in being part of a vital, green environment – it is not considered as a form of memory erasure; it may even be a biological way to stay 'alive' in a manifold form of some kind of natural memory:

How the world changes! And how I change as well!
I am known by one name only, yet

> That which is named by me
> Is not I alone. We are many. I am a living being.
> So that the blood should not freeze in my veins,
> I died many times. Oh, how many dead bodies
> Have I raised from my own body! ...
> I am still alive! More openly, more fully
> Does the spirit embrace the wonderful tribe of creatures.
> Nature is alive. Alive among the rocks
> Is the living grain and my dead herbarium.
> Link into link, form into form. The world
> In all its living architecture is
> A singing organ, a sea of trumpets, a piano
> That does not fade, in joy or when it storms....
> And that which was I, perhaps may
> Grow again and multiply the world of plants....
> Oh, our superstitions!
>
> Zabolotsky 1974 [1937]: 202

Considering how nature can materially hold the remnants of victims could change the field of memory studies in Eastern Europe and its relationship to nature. In the case of events that took place in physical environments in Eastern Europe, history is partly archived in nature. Biological processes transform, digest and conserve historical remains. Thus, the place of memory belongs not only to the cultural realm but also to the environment. However, during the long period of communist propaganda, expressing human traumas under Soviet rule was radically limited, and therefore we still need to wait longer to reread Eastern European environments as more-than-human archives.

Eastern European landscapes, such as forests and the countryside, became the conflicted memorial sites of too many atrocities. They do not represent traditional pastoral nature anymore; they have become disturbed landscapes. Therefore, tensions between cultural and environmental memory exist, but, as the examples of Volhynia and Katyń show, the nationalistic and anthropocentric narrative does not have to be the dominant one. In the absence of human-made memorials, these sites of nature are also sites of memory where nature communicates mourning. Mourning and remembering with nature is part of non-anthropocentric cultural memory, a form of new pastoral and poetic greening of memory. Witnessing in these sites is dispersed among and mediated by exact locations in the environment and human memories. Only through such a relationship can memory be depoliticized and disturbed landscapes speak.

3

Białowieża Forest across Eastern Europe's Borders

> *you have to know that we have with forest*
> *unfinished fears and foggy histories* they have to be told
> Małgorzata Lebda, *The Forest's Border* (2013: 40; my translation)

Looking over a map of Europe from west to east, green forested areas gradually appear as we cross the German–Polish border, and, moving progressively further to the east, there are more and more of them, until we reach the greenest areas, which are located in the European part of Russia. How is this possible, given Soviet Russia's extensive collectivization and hyper-industrialization? In the Soviet era, there are examples of protecting forests and valuable natural areas, such as in the case of the *zapovedniki* (Brain 2012: 227–9), meaning strictly protected reserves or reserves where some basic human management is required. But, more generally, the whole tradition of national parks and reserves in Eastern Europe is rooted in nineteenth-century conservationism and its scientific approach to protecting nature and, at the same time, learning about it. Stalin's forestry politics were contradictory: in some regions, like the Kola peninsula, deforestation accelerated (Josephson et al. 2013: 106), while in others areas, forests were conserved (Brain 2011: 2–3) for the pragmatic purpose of keeping the soil moist to support intensified farming during collectivization. Moreover, forests were extensively planted in so-called green belts during Stalin's Transformation of Nature programme, announced in 1948 after droughts and famine. The plan's main aim was to combat droughts in the steppe and wooded steppe zones of European Russia by planting giant shelter belts and constructing large irrigation channels and water reservoirs widely over three or four years. More than 5,000 kilometres of shelter belts were to be planted in the southern part of the country. In fact, from 1948 to 1953, the number of trees planted exceeded that of those planted during the previous 250 years of forestry history (Josephson et al. 2013: 120). But what kind of forest is spread through these green areas on the map of Russia, and how are they protected now? Not all of the forest is strictly protected in the *zapovedniki* system. These forests are not all ecologically important either: they include patches of young forest, massive industrial pine plantations, weakened forests like the one in Chernobyl that was the first to die out due to the initial wave of poisonous radiation (Mycio 2001: 46).

Let us now turn to the complex case of Białowieża Forest during the bleak period when nature was impatiently sovietized as an economic resource to serve socialism. Białowieża Forest is not just any type of forest: it represents the Eastern European *puszcza* – a thick, old and resilient forest. That does not mean it has not been logged; in fact, deforestation began at the end of the sixteenth century. However, the survival of big mammals (*animalia superiora* including wisents, aurochs, elk, deer and bears) was possible because of a 1577 law that protected the area as royal hunting grounds (Jędrzejewska and Samojlik 2005: 77–8; some historians indicate that the forest was made a hunting ground earlier, in the 1300s, Niklasson et al. 2010: 1320). The widespread exploitation of this forest began in the seventeenth century (80). At the end of that century, Białowieża stopped being protected as royal hunting grounds (82). In the second half of the eighteenth century, the last king of Poland to rescue the collapsing state 'focused on the forest's profitability in the most aggressive capitalistic sense' (84). In 1812, trees were cut down by Napoleon's troops. Serious damage was also inflicted by German occupation during the First World War. Some relief came after Poland regained independence: in 1921 a National Park was established there, but it covered only 3 per cent of the whole area of the forest (Jędrzejewska and Samojlik 2005: 89). Later, Belarusians called this 'the park of virgin nature'. For them, it was a prototype of their national park, Belovezhskaya Pushcha.

After the Second World War, Stalin decided to divide the forest with a border between Poland and the Belarusian Soviet Socialist Republic, and a wide strip of forest was cut down along it. From 1980 on, these two parts of the forest were also separated by a high fence through which neither people nor animals could pass. Large-scale forest clearing was typical for the Soviet period, as American spy satellite images from 1961 document. It is worth noting that before the First World War, stands of trees over a hundred years old occupied about 80 per cent of the area, but by 1958 the share of old stands had fallen to 37.2 per cent, and in 1968, after another ten years of such management, was around 30 per cent (Więcko 1984: 209–12).

In 1944 in Poland, the communist authorities rapidly nationalized most forests, which from that time on were managed by the newly established state forestry commission in Lublin (Konczal 2017: 275). Forestry, as an institution governed by the state, became an instrument for maximizing the production of wood for over fifty years. But local people were also beneficiaries of the Soviet anti-conservationist politics, since 'logging, poaching, or grazing cattle in the woodland' was legal until the 1970s (Blavascunas 2014: 485). In fact, these 'illegal' practices stopped only with the development of tourism in the area, after Poland joined the EU in 2004 (485).

The history of forests in Eastern Europe during the Soviet period spans two extremes: one is a radical instrumentalization of the environment, which started in the Soviet republics in 1918 with the nationalization of forests and the development of forestry into an economic unit; the other is a nationalistic, anthropocentric and symbolic abstraction that impacts the controversial management of the forest in the post-socialist political ecology. In the case of Poland's part of Białowieża Forest, such concepts are driven by patriotic, anti-communist attitudes (i.e. reviving the myth of the so-called 'cursed soldiers') and a prolonged reaction to Soviet colonization that results in a distrust of any foreign influence (i.e. the EU or UNESCO). However, as Eunice Blavascunas noted after many years of ethnographic, anthropological and historical research in the Białowieża forest region: 'post-socialism confused issues of the nation, the periphery, and most importantly, the identity of the Białowieża forest' (2014: 487).

Poland's industrial approach to managing the forest, which includes harvesting timber and culling and hunting wild animals in the oldest parts of the Polish *puszcza*, might be one of the major reasons why even the highly symbolic image of Białowieża – a landscape entangled with national history – is downplayed and manipulated by the state and cannot help to foster the protection of nature. Therefore, we need to revisit the forest's environmental history and find out where the bond between the place of *puszcza* in cultural memory and contemporary forestry has been broken.

What Simon Schama calls the 'long relation between nature and culture' (1995: 10), which is deeply rooted in the history of Central and Eastern European nations and associated with their precious heritage, does not necessarily imply the protection of such landmark natural sites of memory as Białowieża Forest. *Puszcza*, however, represents not only the material forest, but a landscape inscribed in cultural memory by those who have written about it. Material nature cannot be then separated from culture, as Schama argues against this separation in environmental history (12), but can be reconstructed through 'a rich deposit of myths, memories, and obsessions' (14). There are many cultural references to the romanticized image of Białowieża Forest's *puszcza*. However, it is also a disturbed landscape, where pastoral and nationalistic tradition clash with a recent history of mass crimes committed by the Nazis and by the Soviet Secret Police (NKVD).

And it was not just the Soviet period that harmed Białowieża. The forest was extensively logged and devastated during the regimes that followed, and is still being destroyed now. That is why rereading the literary sources related to *puszcza*, in line with Schama's very important thesis that 'landscapes are culture before they are nature' (1995: 61), can draw our attention to the tension between cultural and environmental memory. Schama assumes that there is a consolatory counter-narrative for the history of woodland destruction created by generations of writers, who created the myth of Białowieża landscape. These writers represent pastoral tradition according to which the myth of *puszcza* 'would endure uncontaminated whatever disasters befell the Polish state' (1995: 24). Still, the Soviet period – which was founded on Lenin's materialism, Stalin's environmentally harmful politics 'against nature', and their attempts to demystify and demythicize Eastern European cultures – sheds new light on canonical texts as well as those written during it. Perhaps even the history of Soviet violence against people and environments gave some literary texts new ecocritical meanings. By enabling us to rework the long cultural relation with *puszcza*, they can be read as a form of resistance to Soviet colonization of nature for the whole Eastern European region and its environmental culture. And, as for kings and tsars, the Pushcha again became a hunting ground, now for high-ranking Soviet party officials and the heads of socialist countries. Later, on 8 December 1991, Russian president Boris Yeltsin, Ukrainian president Leonid Kravchuk and the prime minister of Belarus Vyacheslav Kebich signed the agreement to dissolve the USSR at the state dacha near Viskuli in Belovezhskaya Pushcha. This act denounced the 1922 Treaty on the Creation of the USSR and is known today as the Belavezha Accords. The fact that this agreement was signed in the forest strengthened its powerful figurative meaning for Eastern European cultural memory. The collapse of the whole communist bloc, after decades of the Soviet conquest of cultures and natures, happened in this emblematic landscape that survived the Soviet political and environmental colonization.

In what follows, I try to retell this history from the perspective of cultural memory and its relation with the forest, which intersects with environmental history. Informed by Schama's approach, I am, however, more suspicious about what belongs to 'real' national heritage and what has been distorted by nationalistic manipulation of Białowieża Forest's history. For me, it is important to reread the *puszcza* in relation to the complex multi-ethnic history of Eastern European borderlands, since cultural memory of the forest is shared among Poland, Belarus and Lithuania. I would like to emancipate this history from both Soviet and Polish colonial culture (with respect to Poland's colonization of Ukraine, Belarus and Lithuania before the Russian era). My overarching argument is that neither Polish cultural memory nor any other nationalistic and anthropocentric construct of cultural memory protects the forest from being destroyed by humans. These frameworks of culture do not operate in the way they can stop logging. However, there is a missing space between cultural and environmental memory that requires ecocritical rereading of the forest as a text beyond historical Eastern European states' shifting borders.

I will also try to depatriarchize the history of *puszcza*. In Eastern Europe, fathers and sons were involved in military service during the last two world wars, and the forest was often used as a battlefield as well as a place of shelter; it was a soldiers' ally, a partisans' hideout, a hero of songs, a witness to history. This cultural bond with the forest derives from the darkest ancient history of the first legendary settlers in Central and Eastern Europe, and the environmental conditions that they found here. But in contrast to the cultural memory of numerous other countries and great parts of European Russia, Polish militant discourse in particular still ascribes national identity to the forest and includes it in a specific nationalistic version of the history of Poland and the region. The phenomenon of equating Polish national identity with the forest, which partly covers the land area that is now Poland, prompts me to polemicize against this way of narrating history and the state's politics of memory, which lead – against Polish cultural memory, in fact – to legitimization of environmentally controversial operations. I believe the forest has its own history, shaped by literature and art, that will, over the long term, change perceptions of our own situation, our memories, our sociocultural constructions and our political borders. First, though, we need to know the ecological value of this exceptional forest in order to connect its environmental history with cultural memory of it.

Where is Białowieża Forest? What is so special about this environment? The contemporary Białowieża Forest, *Puszcza Bialowieska* in Polish, covers the eastern parts of Poland and the western parts of Belarus (where it is called *Belovezhskaya Pushcha*). More or less 40 per cent of the forest is in Poland, with 60 per cent in Belarus. It is still one of the largest and best-preserved lowland deciduous and mixed forests in Europe. It managed to survive for centuries because, unlike in other European countries where forests were cut for farmland, this forest was protected as a hunting ground for Polish and Lithuanian kings. Its ecosystem and species representation also differ from those of other Eastern European forests, and especially from Russian coniferous and boreal forests (taiga).

Scientists describe Białowieża Forest as 'the last surviving extensive fragment of a primeval temperate forest in the world. It is a relic of vast forests which once extended across the European lowlands and is home to many species that are rare or extinct elsewhere, including wisent (Bison bonasus)', also called European bison, 'and large predators' (Tomiałojć and Wesołowski 2004: 81) such as wolves, lynxes, wildcats, smaller mustelids (badgers, otters, mink, weasels, ermine, fishers, polecats and others), semi-wild horses, raccoon dogs (reintroduced just fifty years ago), boars, foxes, elk, deer, hares, hedgehogs, bats, beavers and numerous birds and rodents. The largest predatory mammal – the brown bear (Ursus arctos) – was extirpated from the forest at the end of the nineteenth century, but still lives in other parts of Eastern Europe (in Baltic countries such as Estonia, Latvia and Russia, and in the Carpathian Mountains) as well as in Western Europe. Attempts to reintroduce brown bears in Białowieża Forest, mainly in the interwar period, were unsuccessful. Despite the fact that the brown bear population is stable according to the IUCN's last assessment (McLellan et al. 2017), scientists indicate that its Eastern European population involves a large group of individuals connected to both the Russian Federation and the Baltic countries. Because these bears relocate across the former Soviet Union's borders, cooperation between these countries (as in the case of Poland and Slovakia) is needed to protect them (McLellan et al. 2017).

Grey wolves (*canis lupus*) were also hunted but managed to survive in the eastern parts of the forest. In 2009, there were three to four packs of wolves that did not have enough room to spread out into, because one pack with a leading male needs around 200 kilometres; similarly, lynxes need up to 300 kilometres. If hunting in the forest were prohibited, the population of large wild predators could increase again (Zub 2009). The protected area of the forest is also too small to host substantial populations of most birds; the numbers of over 70 per cent of species remain below a hundred breeding pairs (Wesołowski et al. 2003). If the transborder post-communist countries cooperated actively, Białowieża Forest could become more rewilded again.

When it comes to trees, there are specific features that make Białowieża Forest typical of rich primeval forests (Tomiałojć and Wesołowski 2004: 82–3):

– Some trees reach a height of over forty or fifty metres and the forest has a multistorey profile;

- A diverse tree community with twenty-six species of trees and fifty-five shrub species. They are also strongly diversified in their age and size. Several individual trees are up to 400 or 500 years old and have the status of national monuments, as indicated by names like Maciek the Oak, or Aleksander Jagiellonczyk, after the king of Poland;
- Large amounts of dead, fallen and uprooted trees.

These characteristics, though, are contested by environmental scientists. Some say that 'primeval' means 'untouched and undisturbed by humans' (Fleming 2002: 1361) and the term as such is overused in the case of Białowieża. They add that, in comparison to European Russian forests, Białowieża is a tiny postage stamp (Fleming 2002: 1361; Szwagrzyk 2016: 292). However, parts of the forest really are protected from human disturbance – these are connected with areas of specialist research, for example on soil (Samojlik et al. 2013).

Critics of the characterization of Białowieża Forest as primeval or untouched by humans are partly right, because only specific fragments of the forest can be materially proven to have such a status. But they also manipulate public opinion when they say that we should forget about essentializing the virginity of this forest due to some old, sentimental dichotomy of man and nature (Szwagrzyk 2016: 292). These scientists justify the foresters' logging by instrumentalizing history to prove how much the forest was dependent on humans and their activity (giving examples of beekeepers, hunters, and the forest as a source of firewood) and emphasize how it has already been destroyed in the past (by fires, ungulates, pests) (Szwagrzyk 2016: 292). There is something really misguided that results particularly from this discourse: it is a distinction between the cultural construction of nature and ecology in forest management. In the case of Białowieża Forest the former does not influence the latter – in other words, the forest's heritage and its potential for conservation collide with a technocratic managerial approach.

The dominant discussion about Białowieża Forest now involves the ongoing problem of logging and other human practices that disturb Białowieża's vulnerable ecosystems and prevent the authorities from extending the borders of national parks, mainly on the Polish side. Despite the fact that UNESCO put Białowieża on their World Heritage Site list in 1979 and the EU designated it as a Natura 2000 Special Area of Conservation, the strictly protected core area covers only 105 km^2 in Poland and 157 km^2 in Belarus. Eighty-three per cent (47,500 hectares) of the forest's core area in Poland is commercially logged and only 17 per cent of it is a national park (10,500 hectares), of which half is strictly protected (Blavascunas 2014: 476). And what is more, Belarus, which is not in the EU, protects the forest more efficiently than Poland, having extended buffer zones around the strictly protected part of the reserve.

In Belarus, the status of the Pushcha was changed in September 1991 by the republic's council of ministers. The hunting farm reserve was transformed into a national park. Its territory was divided into functional zones that made it possible to use not only passive but also active methods of protecting populations of plants and animals. The best-preserved area of old flora in the national park was put on the list of World Heritage Sites in 1992 by UNESCO. Belovezhskaya Pushcha was the first area from the former Soviet Union to be awarded such a prestigious status. In 1993, Belovezhskaya Pushcha was made a biosphere reserve, and at the end of 1997 the Council of Europe awarded it the European Diploma as a benchmark for nature protection institutions on the continent, of which the Belarusians are very proud.

UNESCO also explains on its website that Białowieża belongs to the World Heritage Site list because it represents 'a complex of lowland forests that are characteristic of the Central European mixed forests terrestrial ecoregion. The area has exceptional conservation significance due to the scale of its old growth forests, which include extensive undisturbed areas where natural processes are on-going. A consequence is the richness in dead wood, standing and on the ground, and consequently a high diversity of fungi and saproxylic invertebrates' who dwell on dead wood. 'The property protects a diverse and rich wildlife of which 59 mammal species, over 250 bird, 13 amphibian, 7 reptile and over 12,000 invertebrate species. The iconic symbol of the property is the European Bison: approximately 900 individuals in the whole property which make up almost 25% of the total world's population and over 30% of the free-living animals' (UNESCO).

Against this background, the conflict between Poland under the conservative nationalist Law and Justice Party and international institutions that want to protect the whole of Białowieża Forest is a matter of administrative borders. Jurisdictional boundaries may correspond to ecological ones, as in island nations like Iceland, but not always. 'Physical environments *may* be critical to collective image formation, even – indeed sometimes especially – when imagined territory does not correspond to political unit, as in German Heimat' (Buell 2017: 110). And this is the case with Białowieża Forest and the Polish phantasm of the great Polish nation that existed between the fourteenth and the eighteenth century with its thick *puszcza*. Today the history of Białowieża Forest is reshaped by the current situation, especially the controversy over whether or not to call the whole area of Białowieża a national park. If there are remnants of the primeval forest, then how should they be protected? Or is it enough to argue for forest protection because it represents some primeval forest features? What are the limits of our intervention in the forest's ecosystem and how are they debated?

There are two basic visions of nature that collide with each other in public discourse and predominantly influence the historical narrative about Białowieża: that of the foresters and that of the environmentalists. Foresters in Poland belong to a huge and hierarchically organized company that is owned by the state. There are three CEOs on the national level, with regional directors and 430 forest districts. This is a masculinized group of paramilitary government officials with longstanding patriotic traditions. They wear green uniforms dating back two hundred years with a Polish emblem of the crowned eagle (Konczal 2017: 82). They are managers not only of the forest but also of cultural memory, which is presented in a nationalistic framework: 'the history of the forest becomes the history of Poland and vice versa ... there's no Poland without the forest, and also there's no forestry and no forest and this forestry without Poland' (Konczal 2017: 56). Their work is described as highly moral and devoted to the country, even to God (Konczal 2017: 104), in that Poland and Catholicism have been linked in nationalist culture since the nineteenth century. But in forest management, Polish foresters consider themselves as modern and sustainable, who do not have to protect 'Białowieża as a primeval European relict' (Blavascunas 2014: 479). The importance of Białowieża Forest for the nation is diminished, while the historical role of Polish forester as a genuine patriot is elevated.

Foresters are proud of the 'Polish model of forestry' (Konczal 2017: 119), but they can also be critical of the past. For example, they are aware of the prestige they lost during the Soviet era when the dominant management goal in Białowieża was based on providing wood for industry. This explains why they so strongly accentuate devotion to their homeland and nature, and why they adopt a conservative ethos from pre-Soviet times (especially from the interwar period; Schwartz 2006: 23, 169, in Konczal 2017: 128). The nationalistic attitudes of the foresters and their Polonizing narrative connected with Poland being colonized by neighbouring countries prompt arguments that State Forests should have the right of pre-emption in order to expand the forested areas under state ownership and keep the forest in Polish hands (Konczal 2017: 168–9). Within this patriotic discourse, hunting in the forest is presented as 'an element of a natural environment protection' and a 'persistent part of Polish national culture' (State Forests), but as Olga Tokarczuk shows in her novel *Drive Your Plow Over the Bones of the Dead* (2019 [2009]), it is a highly patriarchal and exclusive activity, which leads to political and legal corruption in the country.

In contrast, we have the side of the environmentalists who concentrate on contemporary problems that arise because of the nationalistic, anti-democratic government of the Law and Justice Party (the party has had a majority in the Polish parliament since the 2015 elections), starting with illegal logging and the culling of bison, elk, boars, moose and wolves. Moreover, the government and foresters promote such specialities as bison meat. Environmentalists have called for help from UNESCO and the EU, with apparent success, but in practice the government does not want to cooperate with international bodies, arguing that forestry has been best developed in Poland – in comparison with other EU countries – and that Poles know better. That is why the state acts as if it does not need to be legitimized by the rest of Europe because it is a sovereign state, sensitive to its postcolonial history (Blavascunas 2014: 480). Environmental activists have tried to represent the forest and its non-human inhabitants' side in this emotionally loaded conflict (a public debate is impossible because neither side can discuss it with the other), but power is with the state, which is undergoing a serious crisis in democracy and ignores European Union directives, including environmental ones.

Rewriting the history of Białowieża means searching for a counter-narrative to the contemporary Polish state politics discourse that has emerged in the wake of Soviet collapse. It is, once again, about emancipating history from the pre-Soviet colonial nostalgia that comes so intensively to the fore in nationalist-dominated post-communist Poland, as well as its post-Soviet distrust of the former colonial state expressed on an international stage. In this history the role of *puszcza* strengthens the emancipatory anti-Soviet discourse across Eastern Europe's borders because of the Soviet period and the replacement of nature with planned socialist ecology that met a different kind of obstacle in Białowieża. Serving as a model of wild nature and pristine forest in environmental cultures and histories of Poland and Lithuania, the forest can be reread as a powerful resource. Its rich deposits of cultural memory contrast with its tragic environmental history.

There are limits to any kind of environmental colonial possession if we dig into the region's cultural memory and recreate the myth of pristine woodland. Try to pronounce the Polish or Belarusian word *puszcza* (pu-shcha), or, in Russian, *пуща*. It is like releasing air, a volatile word, a bird's down. It is etymologically derived from 'emptiness' but it also connotes something fluffy, furry or bushy – something that escapes or is released (from the verb 'puszczać'). Dictionaries detect common synonymous elements in the meaning of *puszcza*, which designates an empty, deserted, unoccupied space, and the word for wilderness in Slavic languages. This is similar to the Hungarian word *puszcza*, meaning 'plain areas', landscape without human communities. But what of the etymology of Bialowieska/Bialowieskaya? It is more human-related. According to Belarusians, the forest acquired its name thanks to a white tower (Belaya Vezha) that was erected more than 700 years ago (between 1276 and 1288) near the small town of Kamianiec. The original name of the tower was Kamyaneckaya Vezha, after the name of the town. The tower (*vezha* in Belarusian) had a strategic military purpose – to watch for the approach of enemy troops. But the word *wieza* belongs to Polish as well, and according to Poles (Długosz 1974 [1455–80]), the building was visited in 1409 by Władysław Jagiełło from Lithuania, who was then the Polish king. We do not know exactly where it was or who erected it, but *puszcza Białowieska* became 'a naturally fortified shelter, where the Polish–Lithuanian nation had begun' (Schama 1995: 61). However, there are other etymological possibilities, for instance that it was white (*biala*) because of snow and because it was occupied by the Jadzwiez or Jadzwingowie tribe, probably of Prussian origins. The Jadzwingowie were the only human tribe that is mentioned in medieval chronicles as living in some parts of Pushcha, but was exterminated at the end of the thirteenth century (Długosz 1974 [1455–80]) – hence the name Białojadzwieska (the old Slavic languages had some vowels that disappeared in later pronunciation). In literature, a romanticized image of the Jadzwiez as a peaceful tribe was created by Jarosław Marek Rymkiewicz in the novel *Polish Conversations in Summer 1983*, which was published as samizdat (in the underground press) in 1984 due to its anti-communist content (2009). Rymkiewicz's book was written during a difficult period of anti-democratic repression, after the Solidarity movement was liquidated and martial law was introduced in 1981. Despite that, the characters discuss Polish identity, spending their holidays at Wigry Lake near Augustów Primeval Forest. But because of their location – the eastern Polish–Lithuanian borderland – the landscape uncovers its centuries-long histories where representatives of other cultures and nations reappear among Poles. The multiple, layered temporal dimensions of this cultural landscape narrative reveal that Polish identity involves a history of multi-ethnic communities, which was typical for all Eastern European borderlands.

The mono-nationalist narrative of Polish history is a nineteenth-century and interwar heritage connected with the country's lost independence; this narrative revived after 1945 in the communist Regained Territories propaganda (see the chapter on mining narratives and Silesia), and even more intensively after the collapse of the Eastern bloc in 1991. Norman Davies, in an interview, said that it is unnatural for contemporary Poland to be so homogenous given its multicultural past, and with this past young Poles identify more (Davies, Youtube.com). And I think counteracting the nationalist myth suggests a role for the forest that it has inhabited in Eastern European cultural memory.

Białowieża Forest represents one such Eastern European landscape that belongs not only to Polish culture but also to Lithuanian, Belarusian and Russian culture, and beyond – it is a former lowland landscape that stretched across the European Plain after the last glacial period before agricultural human societies developed. It is a landscape historically encircled by collectives of people who spoke not just Polish and Belarusian. Their mostly Slavic ancestors belonged to a vivid multilingual, multinational, multireligious Eastern European community that constituted the Polish–Lithuanian Commonwealth until 1795. There one could hear, apart from Polish, also Ruthenian (early Belarusian), Lithuanian, German, Hebrew, Yiddish, Italian, Armenian, Arabic and French. In fact, Białowieża is commonly associated with a text of identity culture by the prophetic, romantic Polish–Lithuanian poet Adam Mickiewicz, *Pan Tadeusz or the Last Foray in Lithuania: A Story of Life among Polish Gentlefolk in the Years 1811 and 1812 in Twelve Books*. This epic was written in 1834 and all Poles are required to read it in school (and were so even during the communist era – I had to learn some parts of it by heart). But Poles are rarely taught that this text is one of the great signs of environmental consciousness expressed in national literature. Book Four includes a fragment of a description of the deepest heart of the forest, which gathers and preserves all species and which humans cannot enter because it is a woodland's lush and hostile wetland, with swarms of insects and deep swamps, that is impossible to traverse:

> But finally behind this mist (so runs the common rumour) extends a very fair and fertile region, the main capital of the kingdom of beasts and plants. In it are gathered the seeds of all trees and herbs, from which their varieties spread abroad throughout the world; in it, as in Noah's ark, of all the kinds of beasts there is preserved at least one pair for breeding. In the very centre, we are told, the ancient buffalo and the bison and the bear, the emperors of the forest, hold their court. Around them, on trees, nest the swift lynx and the greedy wolverene, as watchful ministers; but farther on, as subordinate, noble vassals, dwell wild boars, wolves, and horned elks. Above their heads are the falcons and wild eagles, who live from the lords' tables, as court parasites. These chief and patriarchal pairs of beasts, hidden in the kernel of the forest, invisible to the world, send their children beyond the confines of the wood as colonists, but themselves in their capital enjoy repose; they never perish by cut or by shot, but when old die

by a natural death. They have likewise their graveyard, where, when near to death, the birds lay their feathers and the quadrupeds their fur. The bear, when with his blunted teeth he cannot chew his food; the decrepit stag, when he can scarcely move his legs; the venerable hare, when his blood already thickens in his veins; the raven, when he grows grey, and the falcon, when he grows blind; the eagle, when his old beak is bent into such a bow that it is shut for ever and provides no nourishment for his throat; all go to the graveyard. Even a lesser beast, when wounded or sick, runs to die in the land of its fathers. Hence in the accessible places, to which man resorts, there are never found the bones of dead animals. It is said that there in the capital the beasts lead a well-ordered life, for they govern themselves; not yet corrupted by human civilisation, they know no rights of property, which embroil our world; they know neither duels nor the art of war. . . . Happily no man wanders into this enclosure, for Toil and Terror and Death forbid him access.

1917 [1834]: 105–6

This so-called 'motherland', not fatherland, 'of woods' resembles a strict nature reserve, the mythical ark where animals, free from the threat of humans, come to die. In the 'kernel' of the forest, in its wildest, thick parts, animals are protected from human access. The half-legendary, half-mythical – 'so runs the common rumour' – status of this Romantic pastoral image of unspoiled nature discloses uncorrupted animal 'capital' and hierarchy of species there in accordance with cultural recognition that some types of animals are more royal and noble than the others. 'The emperors of the forest' – the extinct aurochs and unseen today brown bear – used to have the habitat in this forest; the bison was successfully reintroduced thanks to its iconic character for this landscape almost one hundred years later than Mickiewicz's text. And, in the middle of this forest, animals have their invisible for people graveyard, valued by their right to natural death (when they are old and sick). Mickiewicz's image of nature's sanctuary forms a powerful conservationist narrative, but the legendary image of the animal cemetery ('the bones of dead animals' cannot be found) links this natural landscape with cultural memory in a special way. We cannot rescue and protect all species but we can cherish the memory of animal ancestors by recalling such images as Mickiewicz's heart of the woodland. Environmental cultures can be fuelled by these imaginary tropes of ecocentric memory: to remember the mythical *puszcza* despite the environmental history of logging, poaching and hunting.

Schama devotes a long passage to Mickiewicz, who 'had described the etiology of the ancient forest with such a keen eye' (1995: 58). For Mickiewicz, who imagines Białowieża on emigration in Paris, it is a natural landscape of memory, a romantic construct of imagination (60–1). But this is also a text of resistance: in Mickiewicz's time, it protested against Russian colonization; reread during the communist period, it protested against sovietization. Its rebellious character is expressed in part through the description of a world free from human violence in the heart of *puszcza*. It is also expressed through the forest's environmental history, as an invocation to the oldest trees represents the cultural memory of these lands (see Figure 11):

> Monuments of our fathers! how many of you each year are destroyed by the axes of the merchants, or of the Muscovite government! These vandals leave no refuge either for the forest warblers or for the bards, to whom your shade was as dear as to the birds. Yet the linden of Czarnolas, responsive to the voice of Jan Kochanowski, inspired in him so many rimes! Yet that prattling oak still sings of so many marvels to the Cossack bard!
>
> How much do I owe to you, trees of my Homeland! A wretched huntsman, fleeing from the mockery of my comrades, in exchange for the game that I missed how many fancies did I capture beneath your calm, when in the wild thicket, forgetful of the chase, I sat me down amid a clump of trees!
>
> 1917 [1834]: 92

Figure 11 Białowieża Forest. Photo by Benno Hansen, Foter.com. CC BY

Mickiewicz was one of the writers who filled this historically multilayered landscape with environmental values. *Puszcza* was a channel for him to communicate the idea of national emancipation because he knew how to read this landscape despite having emigrated. However, he ecologized it first, rather than putting it in a nationalistic corset.

Schama is right that Białowieża became a 'patriotic landscape' due to its cultural resources: in Artur Grottger's drawings or Stefan Żeromski's writings (1995: 62–3) the power of the *puszcza* voice always accompanied the expression of human trauma. The extreme crimes committed by the Nazi and Soviet regimes against Jews and Poles in the heart of Białowieża Forest also disturbed the forest's pastoral voice, which has been transcribed by the sensitive radar of literature during the communist repressions.

The Soviet period of extensive logging muffled the *puszcza*. The Soviet and Polish parts of Białowieża started to blur together problematically in cultural memory as well, and the Romantic image of the pristine forest started to vanish. Some writers, such as Igor Newerly, nostalgically expressed the emptiness in memory when the forest lost its cultural meaning due to physical environmental harm (2018 [1986]). But a more disturbing voice about the *puszcza*'s ambivalent role in cultural memory during the hopeless communist period came from the writer Tadeusz Konwicki, like Mickiewicz of Lithuanian origins. In *Kompleks Polski* ('Polish Complex'), first published in the underground in 1977, he borrowed a pastoral myth of the Białowieża and the lyricism of romantic patriotic writers, but he also transformed these according to his experience of the bleak reality of communism. The primeval *puszcza*'s state, where freedom is embodied by a forest animal, is also seen as indifferent to historical events because only here do Polish independence uprisings from the nineteenth century mingle with totalitarian atrocities. That is why Konwicki's narrator is lost in the forest landscape, which became a hostile battlefield, a disturbed landscape of human history:

> More and more soldiers entered the barrage. They were gathering here in groups. They discussed something, sometimes one of them shot into the forest, the contemptuously silent forest, indifferent, was melting above in the rain. (1990 [1977]: 59; my translation)

> There were neither insurgents nor Muscovites. There was only a busy forest, a forest roaring like a mill, there was rain and there was a sense of passivity in the autumnal trees' needles. You started running on the rough terrain around the road. Everywhere lay abandoned weapons: shotguns, scabbards, revolvers, patrons. Next to the fallen tree you saw a banner, next to it was a shotgun

> dropped while fleeing. They all ran away from one shot. Both attacked and harassing.
>
> ... You were stumbling, tripping over creepy roots, these fossilized varicose veins of the forest.
>
> 60–2

Being lost in a history of nineteenth-century insurgencies that intersects with communist oppression revives the multilayered historical memory of the *puszcza* as a witness to past events in Konwicki's text. Perhaps seen as indifferent, the *puszcza* also reveals something else about the period of hostile colonization. What does it mean to be trapped? Animal fear did not accompany the emigrant Mickiewicz's nostalgic vision, but Konwicki's protagonists, who were thrown into communism just after war executions and genocide, inscribed a different experience in the forest. The chaos of history is portrayed through scattered gunshots in the disturbed landscape of *puszcza* and the paradox of perceiving the forest as 'busy' separately from human history, 'roaring' in its own voice. However, when the protagonist is crashing through the forest, he recognizes 'the fossilized varicose veins' of the trees' roots as obstacles. They symbolize the inevitable, but painful, bond between human and environmental history.

Under communism, the forest became a 'green Babel tower' (Soroczyński 1978: 208), divided between human traumatic appropriations. However, still able to fill cultural memory with the uncontaminated language of romanticized spirituality, which was particularly needed given the phantasmic reality of Soviet propaganda. In 1975, Stanisław Grochowiak wrote a prayer to Białowieża Forest titled *Antiphon* and restored the myth of the *puszcza* as a soothing landscape for a politically distorted society:

> Bialowieża Blue
> Grass of Void Meadows
> Wisdom shooting range
> For the right purpose
> Save human torment
> Don't expel us from animals
> ...
> Don't condemn liars
> Straighten them with the hazel
> Wash us with a forest stream
> Too eager smear
> Bialowieża Blue
> Red rudiment of oaks

Show
Even to stubborn blind men

Spruce on the throne
Underwater glade
Where hair is dissolved by lush greenery
Where the great diving beetle circles above its thicket
And the sun thinly shines in the eclipse

Keep us in mind.

2001; my translation

The poem's title derives from the Greek terms 'in return' and 'sound', and known in Christian liturgy as a verse commencing or finishing a prayer like a biblical psalm or canticle. Białowieża is sanctified with the colour blue, called on in a search for righteous wisdom. The forest is asked to shelter human animals, and believed that she can purify their mistakes. If this is the heart of thick *puszcza* signified by 'red oaks' and 'lush greenery', then, like Mickiewicz's motherland, it is a powerful figure of cultural memory.

The poem cannot be inscribed into any nationalistic ideology. During a period of deep political disappointment, it reinserts the pastoral myth in place of a disturbed landscape of memory in the hope – though 'the sun thinly shines in the eclipse' – that the Soviet regime can be abolished. In the 1970s when Grochowiak wrote this poem, nobody believed that it would soon happen. But the question is, will the forest survive as more than a cultural construct of memory?

What does it mean to preserve the forest? Stopping it at a particular moment in its continual change? Achieving a state of apparent 'wildness' or maximizing its biodiversity? We are dealing with a type of forest that started developing after the last ice age more than 10,000 years ago, when there was also a biologically rich area of mixed boreal forest with tundra, where such giant creatures as woolly mammoths and rhinoceroses, steppe bison, musk oxen, cave bears, lions, giant elk and straight-tusked elephants ran. The historical existence of the forest we know is just part of a larger world beyond humans, full of beasts and fascinating ancient creatures. This environmental history does not infiltrate cultural memory.

Nature participates in Eastern European history otherwise, through the socioculturally relevant context of memorialization. The forests, including Białowieża, were severely harmed during the communist period as the rest of the environment: 'destroyed by acid rain, soils were contaminated by heavy metals, polluting industries belched dioxins into the sky, and the Chernobyl nuclear disaster occurred in 1986' (Blavascunas 2014: 479). And they survived even Soviet colonization of Eastern European nature. But without literature, the environmental history of Białowieża Forest is in a way anti-environmental. It can be summarized by a simple yet sad description, posted on the official website of the Belarusian Belovezhskaya Pushcha National Park: 'The territory of Belovezhskaya Pushcha was passed from one state to another time and again, but almost always it was a hunting place for the highest dignitaries. Since time immemorial, Kievan and Lithuanian princes, Polish kings, Russian tsars, general secretaries hunted here, saving it for their hunting' (the National Park 'Belovezhskaya Pushcha').

In cultural memory, the landscape of Białowieża Forest fuels the narration of human history, no matter whether it is Polish, Belarusian or European, anthropocentric or ecocentric. To rescue the forest is to try to protect it in a meaningful entanglement in memory as a potential source of political resistance, as well as an open and alive reference for Eastern European environmental culture. The Białowieża forest's 'memory of sylvan virtue' not only belongs to Polish–Lithuanian tradition and protects the 'hidden heart of national identity', as Schama writes (56), but also becomes a source of anti-communist voice across Eastern Europe's borders. It is a special landscape because what human history has inscribed in this forest, and how violence disturbed it, cannot be forgotten due to the powerful idea of the Romantic and pastoral myth of *puszcza*. This image turned to be readable and interpreted due to the course of Eastern European history and in resistance to Soviet cultural and environmental colonization. It took different variants but never has been fully rejected. Can it also serve as an argument for the forest's ecological protection?

* * *

A famous Soviet music band from Belarus called *Pesniary* created a moving tribute to the forest, which was especially popular in the 1970s. It is a beautifully melancholic song, composed in 1975 by Aleksandra Pakhmutova with lyrics by Nikolai Dobronravov, representing this unspoiled romanticized myth of a forest landscape, tenderly preserved in the cultural memory of Soviet Eastern Europe:

> Treasured chant, treasured vast distance
> Crystal light of the dawn, rising up over the world.
> Your endless sadness is so clear for me,
> Belovezhskaya pushcha, Belovezhskaya pushcha

There is a long forgotten home we all came from.
And when voice of ancestors calls me sometimes
I'm like a grey forest bird who is flying to you,
Belovezhskaya pushcha, Belovezhskaya pushcha

By barely visible trail I'm reaching the creek,
Where the grass is so high, where the undergrowth is so dense.
Like the deer on their knees, I drink thy holy spring truth,
Belovezhskaya pushcha

At high birches, getting warmhearted,
I will take away to comfort those who stay
Your treasured chant, your miraculous melody,
Belovezhskaya pushcha, Belovezhskaya pushcha

      translated by Mikhail Palstsianau, lyrictranslate.com

Białowieża is a cultural treasure across Eastern European borders, a 'long forgotten home', which was distorted during the Soviet period. However, it has survived this contaminated world: its contaminated by propaganda language, landscape, memory and even history. It has survived the time when the ties between human and natural worlds were broken, and when this broken world turned to be more wounded than revolutionary. The forest, perhaps more than any other Eastern European landscape, is a polyphonic text of this region's environmental culture where voice of human and non-human 'ancestors call' with its 'endless sadness'. It is striking, though, that the memory of Eastern Europe, soaked in so much human death and ache and real blood, is still able to recreate this romanticized myth of nature.

Notes

1. In Russian, *istóriya* means both 'history' and 'a story', the same as in other Slavic languages.
2. In Daniel Weissbort's translation, the ending of the poem *Yesterday, as I thought about Death* is different: 'I myself was not a child of nature, / but her thought, her inconstant mind!' (Zabolotsky 1999: 156).
3. *Konopielka* was also adapted as a film by Witold Leszczyński (1981). This film, as well as the novel, reveals East Poland's rural landscape and documents an original Eastern Polish rural dialect.
4. All translations are mine. I have not rendered the rural dialect, since it is hard to understand even in Polish. I think this is one of the reasons why this famous Polish peasant novel has not yet been translated into other languages.
5. However, the controversial construction of the Siemianówka Dam Reservoir after 1977 distorted the ecosystem of the Narew basin (Grabowska, Ejsmont-Karabin and Karpowicz 2013).
6. This unit was mixed up with the rem (or roentgens, after Wilhelm Roentgen) during the Chernobyl catastrophe, causing confusion.
7. Ursula Heise also mentions the play *Sarcophagus: A Tragedy* by Vladimir Gubaryev, and the novels *The Star Chernobyl* by Julia Voznesenskaya and *Chernobyl* by Frederik Pohl, all released in 1987 (2008: 180).
8. The term 'disturbed landscapes' is borrowed from Peter Bock-Schroeder, who was the first West German photographer to get permission to work in the USSR. In the text from 1964, which is included in his photographical series of the same title, the artist explains his fascination with landscapes altered by humans in contrast to 'sterile' natural landscapes.
9. Karolina Grzywnowicz – in her artistic project *The Weeds*, realized in 2015 in Warsaw gallery Zachęta – evoked those places of deportations from the eastern borderlands of Poland that could be identified by the species of plants that were growing in common gardens.
10. The concept of 'carriers of memory' was created by Polish historian Marcin Kula (2002), who includes elements of nature like trees, and serves to indicate all material and immaterial aspects of stimulating memory (memory triggers).

Bibliography

Alaimo, Stacy. 'Elemental Love in the Anthropocene'. In *Elemental Ecocriticism. Thinking with Earth, Air, Water, and Fire*, edited by Jeffrey Jerome Cohen and Lowell Duckert, 298–309. Minneapolis: University of Minnesota Press, 2015.

Alexievich, Svetlana. *Chernobyl Prayer. A Chronicle of the Future*. Translated by Anna Gunin, Arch Tait. London: Penguin Random House, 2016.

Andruchovych, Jurij and Andrzej Stasiuk. *Moja Europa. Dwa eseje o Europie zwanej Środkową*. Wołowiec: Czarne, 2000.

Angus, Jan. *Facing the Anthropocene. Fossil Capitalism and the Crisis of the Earth System.* New York: Monthly Review Press, 2016.

Ankersmit, Frank. *History and Tropology: The Rise and Fall of Metaphor*. Berkeley: University of California Press, 1994.

Applebaum, Anne. *Gulag. A History of the Soviet Prison Camp*. New York: Doubleday, 2003.

Ash, Timothy Garton. *The Uses of Adversity: Essays on the Fate of Central Europe*. Cambridge: Granta, 1989.

Assmann, Aleida. *Cultural Memory and Western Civilization: Functions, Media, Archives*. Cambridge: Cambridge University Press, 2012.

Assmann, Jan. *Religion and Cultural Memory: Ten Studies*. Stanford: Stanford University Press, 2006.

Assmann, Jan. *Cultural Memory and Early Civilisation. Writing, Remembrance, and Political Imagination*. New York: Cambridge University Press, 2011.

Baer, Ulrich. *Spectral Evidence: The Photography of Trauma*. Cambridge MA: MIT Press, 2002.

Bajcer, Stanisław. 'Rok 1944. Pieśń patriotyczna Narodowych Sił'. *Zeszyty do historii Narodowych Sił Zbrojnych* V (1990): 135.

Balassa, Bela A. 'Collectivization in Hungarian Agriculture'. *Journal of Farm Economics* 42, no. 1 (1960): 35–51.

Bałka, Mirosław. 'Amidst Shadows…and Little Roes'. Rafał Jakubowicz talks to Mirosław Bałka. In *Winterreise. The exhibition's catalogue*. Kraków: Galeria Starmach, 2004.

Baratay, Éric. *Bêtes des tranchées. Des vécus oubliés*. Seuil: CNRS, 2013.

Barcz, Anna, Petra Buchta-Bartodziej and Anna Michalak. 'The Oder – a River that Floods: The Problem of Environmental Adaptation in Literary Texts'. *Environmental Hazards* 17, no. 3 (2018): 251–67.

Bartoszyński, Kazimierz. 'Aspekty i relacje tekstów (Źródło – historia – literatura)'. In *Dzieło literackie jako źródło historyczne,* edited by Zofia Stefanowska and Janusz Sławiński, 52–93. Warszawa: Czytelnik, 1978.

Batori, Anna. *Space in Romanian and Hungarian Cinema*. Cham: Palgrave Macmillan, 2018.

Bauman, Zygmunt. *Liquid Modernity*. Cambridge: Polity Press, 2000.

Beck, Ulrich. *World at Risk*. Translated by Ciaran Cronin. Cambridge: Polity Press, 2009.

Bell, Peter D. *Peasants in Socialist Transition. Life in a Collectivized Hungarian Village*. Berkeley: University of California Press, 1984.

Belov, Fedor. *The History of a Soviet Collective Farm*. London: Routledge, 1998.

Belyaev, Sergey and Boris Pilnyak. *Meat: A Novel*. Translated by Ronald. D. LeBlanc. University of New Hampshire, Faculty Publications, 2019 [1936].

Bergthaller, Hannes, Rob Emmett, Adeline Johns-Putra, Susanna Lidström, Agnes Limmer, Shane McCorristine, Isabel Pérez Ramos, Dana Phillips, Kate Rigby and Libby Robin. 'Mapping Common Ground: Ecocriticism, Environmental History, and the Environmental Humanities'. *Environmental Humanities* 5 (2014): 261–76.

Blackbourn, David. *The Conquest of Nature. Water, Landscape, and the Making of Modern Germany*. London-New York: W. W. Norton, 2006.

Blacker, Uilleam. 'The Wood Comes to Dunsinane Hill: Representations of the Katyn Massacre in Polish Literature'. *Central Europe* 10, no. 2 (2012): 108–23.

Blavascunas, Eunice. 'When foresters reterritorialize the periphery: post-socialist forest politics in Białowieża, Poland'. *Journal of Political Ecology* 21, no. 1 (2014): 475–92.

Borzęcki, Robert. 'Górnictwo uranu w Polce'. *Otoczak*, no. 31 (2004): 28–43.

Brain, Stephen. *Song of the Forest. Russian Forestry and Stalinist Environmentalism, 1905–1953*. Pittsburgh: University of Pittsburgh Press, 2011.

Brain, Stephen. 'The Environmental History of the Soviet Union'. In *A Companion to Global Environmental History*, edited by J. R. McNeill and Erin Stewart Mauldin, 222–43. Chichester: Blackwell Publishing, 2012.

Bramwell, Anna. *Ecology in the Twentieth Century: A History*. New Haven: Yale University Press, 1989.

Breyfogle, Nicholas B. (ed.). *Eurasian Environments. Nature and Ecology in Imperial Russian and Soviet History*. Pittsburgh: University of Pittsburgh Press, 2018.

Brock, Emily K. 'New Patterns in Old Places: Forest History for the Global Present'. In *The Oxford Handbook of Environmental History*, edited by Andrew C. Isenberg, 154–77. Oxford: Oxford University Press, 2017.

Brodsky, Joseph. *Selected Poems*. Translated by George L. Kline. Harmondsworth: Penguin Books, 1973.

Browarny, Wojciech. *Tadeusz Różewicz and Modern Identity in Poland since the Second World War*. Wrocław: Fundacja na rzecz promocji nauki polskiej, 2019.

Brown, Deming. *Soviet Russian Literature since Stalin*. Cambridge: Cambridge University Press, 1978.

Brown, Kate. *Plutopia. Nuclear Families, Atomic Cities, and the Great Soviet and American Plutonium Disasters*. Oxford: Oxford University Press, 2013.

Brown, Kate. *Manual for Survival. A Chernobyl Guide to Survival*. London: W. W. Norton & Company. 2019.

Bruno, Andy. *The Nature of Soviet Power: An Arctic Environmental History*. Cambridge: Cambridge University Press, 2016.

Bruno, Andy. 'How a Rock Remade the Soviet North. Nepheline in the Khibiny Mountains'. In *Eurasian Environments. Nature and Ecology in Imperial Russian and Soviet History,* edited by Nicholas B. Breyfogle, 147–64. Pittsburgh: University of Pittsburgh Press, 2018.

Buell, Lawrence. 'Toxic Discourse'. *Critical Inquiry* 24, no. 3 (1998): 639–65.

Buell, Lawrence. 'Uses and Abuses of Environmental Memory'. In *Contesting Environmental Imaginaries. Nature and Counternature in a Time of Global Change*, edited by Steven Hartman, 95–116. Leiden: Brill, 2017.

Çağlayan, Emre. *Poetics of Slow Cinema. Nostalgia, Absurdism, Boredom*. Cham: Palgrave Macmillan, 2018.

Carey, Mark. 'Beyond Weather: The Culture and Politics of Climate History'. In *The Oxford Handbook of Environmental History*, edited by Andrew C. Isenberg, 23–51. Oxford: Oxford University Press, 2017.

Carter, Francis W. and David Turnock (eds). *Environmental Problems in Eastern Europe*. London: Routledge, 1996.

Chakrabarty, Dipesh. 'The Climate of History: Four Theses'. *Critical Inquiry* no. 35 (2008): 197–222.

Charlesworth, Andrew. 'The Topography of Genocide'. In *The Historiography of the Holocaust*, edited by Dan Stone, 216–52. Basingstoke: Palgrave Macmillan, 2004.

Cheloukhina, Svetlana. '"Torzhestvo zemledeliia": One More Time on the Famous Poèma by Zabolotsky'. In *Sbornik, posviashchennyii 110-letiiu so dnia rozhdeniia N.A. Zabolotskogo*, 185–206. Moscow: Azbukovnik, 2013.

Chukhontsev, Oleg. 'When the Village Daylight Dimmed'. Translated by Simon Franklin. In *Twentieth Century Russian Poetry. Silver and Steel. An Anthology*, edited by Yevgeny Yevtushenko, Albert Todd and Max Hayward, 952–53. New York: Doubleday, 1993 [1968].

Clark, Katerina. 'The Centrality of Rural Themes in Postwar Soviet Fiction'. In *Perspectives on Literature and Society in Eastern and Western Europe*, edited by Geoffrey A. Hosking and George F. Cushing, 76–100. London: Palgrave Macmillan, 1989.

Clark, Timothy. *Ecocriticism on the Edge. The Anthropocene as a Threshold Concept*. London: Bloomsbury Academic, 2015.

Cohen, Jeffrey Jerome and Lowell Duckert. 'Eleven Principles of Elements'. In *Elemental Ecocriticism. Thinking with Earth, Air, Water, and Fire,* edited by Jeffrey Jerome Cohen and Lowell Duckert, 1–26. Minneapolis-London: University of Minnesota Press, 2015.

Cole, Tim. '"Nature was helping us." Forests, Trees and Environmental Histories of the Holocaust'. *Environmental History* 19 (2014): 665–86.

Collingwood, R. G. *The Idea of History*. Oxford: Clarendon Press, 1946.
Czapliński, Przemysław. 'Historia, narracja, sprawczość'. In *Historia – dziś. Teoretyczne problemy wiedzy o przeszłości*, edited by Ewa Domańska, Rafał Stobiecki and Tomasz Wiślicz, 283–301. Kraków: Universitas, 2014.
Czapliński, Przemysław. 'Poszerzanie pola Zagłady'. *Teksty Drugie* 164, no. 2 (2017): 7–16.
Czyżak, Agnieszka. 'Śląskie lochy i smoki – wariacje historyczne Szczepana Twardocha'. *Czytanie Literatury. Łódzkie Studia Literaturoznawcze* 5 (2016): 37–45.
Danta, Darrick. 'Great Hungarian Plain'. In *Encyclopedia of Eastern Europe: From the Congress of Vienna to the Fall of Communism*, edited by Richard Frucht, 261. New York: Garland Publishing, 2000.
De Luca, Tiago and Nuno Barradas (eds). *Slow Cinema*. Edinburgh: Edinburgh University Press, 2016.
Długosz, Jan. *Roczniki, czyli kroniki sławnego Królestwa Polskiego* ks. VII–VIII. Translated from Latin by Julia Mruwkówna, edited by Danuta Turkowska and Maria Kowalczyk. Warszawa: PWN, 1974.
Domańska, Ewa. 'The Material Presence of the Past'. Translated by Magdalena Zapedowska. *History and Theory* 45 (2006): 337–48.
Domańska, Ewa. 'Przestrzenie Zagłady w perspektywie ekologiczno-nekrologicznej'. *Teksty Drugie* 164, no. 2 (2017): 34–60.
Duckert, Lowell. 'When It Rains'. In *Material Ecocriticism*, edited by Serenella Iovino and Serpil Oppermann, 114–29. Bloomington: Indiana University Press, 2014.
Duckert, Lowell. 'Earth's Prospects'. In *Elemental Ecocriticism. Thinking with Earth, Air, Water, and Fire*, edited by Jeffrey Jerome Cohen and Lowell Duckert, 237–67. Minneapolis: University of Minnesota Press, 2015.
Errl, Astrid. *Kollektives Gedächtnis und Erinnerungskulturen*. Stuttgart-Weimar: J.B. Metzler, 2005.
Erll, Astrid and Ansgar Nünning (eds). *Cultural Memory Studies: An International and Interdisciplinary Handbook*. New York-Berlin: De Gruyter, 2008.
Erll, Astrid. *Memory in Culture*. Translated by Sara B. Young. New York: Palgrave Macmillan, 2011.
Fenyo, Mario D. 'Hungary (History)'. In *Encyclopedia of Eastern Europe: From the Congress of Vienna to the Fall of Communism*, edited by Richard Frucht, 296–315. New York: Garland Publishing, 2000.
Feshbach, Murray and Alfred Friendly. *Ecocide in the USSR: Health and Nature under Siege*. New York: Basic Books, 1992.
Feshbach, Murray. *Ecological Disaster: Cleaning up the Hidden Legacy of the Soviet Regime*. New York: Twentieth Century Fund Press, 1995.
Fitzpatrick, Sheila. *Stalin's Peasants. Resistance and Survival in the Russian Village after Collectivization*. New York: Oxford University Press, 1994.
Fiut, Aleksander. *The Eternal Moment. The Poetry of Czesław Miłosz*. Translated by Theodosia S. Robertson. Berkeley: University of California Press, 1990.

Fleming, Michael. 'The Bialowieza Forest Saga by Simona Kossak. Review'. *Europe-Asia Studies* 54, no. 8 (2002): 1361–3.
Frydryczak, Barbara. 'Krajobraz'. In *Modi Memorandi. Leksykon kultury pamięci*, edited by Magdalena Saryusz-Wolska and Robert Traba, 195–8. Warszawa: Scholar, 2014.
Garrard, Greg, Gary Handwerk and Sabine Wilke. 'Imagining Anew: Challenges of Representing the Anthropocene'. *Environmental Humanities* 5 (2014): 149–53.
Gifford, Terry. *Pastoral*. London: Routledge, 1999.
Głowiński, Michał. *Gry powieściowe*. Warszawa: PWN, 1973.
Głowiński, Michał. 'Lektura dzieła a wiedza historyczna'. In *Dzieło literackie jako źródło historyczne*, edited by Zofia Stefanowska and Janusz Sławiński, 52–93. Warszawa: Czytelnik, 1978.
Goldstein, Darra. *Nikolai Zabolotsky. Play for Mortal Stakes*. Cambridge: Cambridge University Press, 1993.
Grabowska, Magdalena, Jolanta Ejsmont-Karabin and Maciej Karpowicz. 'Reservoir-river relationships in lowland, shallow, eutrophic systems: an impact of zooplankton from hypertrophic reservoir on river zooplankton'. *Polish Journal of Ecology* 61, no. 4 (2013): 759–68.
Greg, Wioletta. *Swallowing Mercury*. Translated by Eliza Marciniak. London: Portobello Books, 2017.
Grewe, Bernd-Stefan. 'Forest History'. In *The Turning Points of Environmental History*, edited by Frank Uekoetter, 44–54. Pittsburgh: University of Pittsburgh Press, 2010.
Grochowiak, Stanisław. 'Antyfona'. In *Poezje wybrane*. Warszawa: PIW, 2001 [1975].
Grossman, Joan Delaney. 'The Transformation Myth in Russian Modernism: Ivan Konevskoi and Nikolai Zabolotsky'. In *Metamorphoses of Russian Modernism*, edited by Peter I. Barta, 41–60. Budapest: Central European University Press, 2000.
Györffy, M. *A tizedik évtized. A Magyar játékfilm a kilencvenes években és más tanulmányok* [Hungarian film in the 1990s]. Budapest: Palatinus Kiadó, 2001.
Halbwachs, Maurice. *Les cadres sociaux de la mémoire*. Paris: Librairie Félix Alcan, 1925.
Halbwachs, Maurice. *La Topographie légendaire des Évangiles en Terre Sainte. Étude de mémoire collective*. Paris: Presses Universitaires de France, 1941.
Halbwachs, Maurice. *La Mémoire collective*. Paris: Albin Michel, 1950. [*On Collective Memory*]. Translated by Lewis A. Coser. Chicago-London: The University of Chicago Press, 1992.
Hartwig, Julia. *Chwila postoju*. Kraków: Wydawnictwo Literackie, 1980.
Hartwig, Julia. *Jasne niejasne*. Kraków: Wyd. a5, 2009.
Heise, Ursula K. *Sense of Place and Sense of Planet. The Environmental Imagination of the Global*. New York: Oxford University Press, 2008.
Heller, Mikhail and Aleksandr Nekrich. *Utopia in Power. The History of the Soviet Union from 1917 to the Present*. Translated by Phyllis B. Carlos. London: Hutchinson, 1986.
Hosking, Geoffrey. *A History of the Soviet Union*. London: Fontana Press/Collins, 1985.

Hosking, Geoffrey. 'The Institutionalisation of Soviet Literature'. In *Perspectives on Literature and Society in Eastern and Western Europe*, edited by Geoffrey A. Hosking and George F. Cushing, 55–75. London: Palgrave Macmillan, 1989.

Hölderlin, Friedrich. *Hyperion and Selected Poems*, edited by Eric L. Santner. New York: Continuum, 1990.

Huk, Bogdan. *Ukraina. Polskie jądro ciemności*. Przemyśl: Stowarzyszenie Ukraińskie Dziedzictwo, 2013.

Hundorova, Tamara. 'Czarnobyl, nuklearna apokalipsa i postmodernism'. *Teksty Drugie* 6 (2014): 249–63.

Hundorova, Tamara. 'Gatunek czarnobylski: wyparcie realnego i nuklearna sublimacja'. Translated by Przemysław Tomanek. In *Po Czarnobylu. Miejsce katastrofy w dyskursie współczesnej humanistyki*, edited by Iwona Boruszkowska, Katarzyna Glinianowicz, Aleksandra Grzemska and Paweł Krupa, 55–66. Kraków: Wydawnictwo Uniwersytetu Jagiellońskiego, 2017.

Illyés, Gyula. 'The Three Old Stagers' [an excerpt]. Translated by Michael Beevor. In *Selected Poems*, edited by Thomas Kabdebo and Paul Tabori. London: Chatto & Windus, 1971 [1932].

Isenberg, Andrew C. *The Oxford Handbook of Environmental History*. Oxford: Oxford University Press, 2017.

Jaffe, Ira. *Slow Movies: Countering the Cinema of Action*. London: Wallflower Press, 2014.

Jaros, Jerzy. *Historia górnictwa węglowego w Polsce Ludowej (1945–1970)*. Warszawa-Kraków: PWN, 1973.

Jarosz, Dariusz. 'The Collectivization of Agriculture in Poland. Causes of Defeat'. In *The Collectivization of Agriculture in Communist Eastern Europe: Comparison and Entanglements*, edited by Constantin Iordachi and Arnd Bauerkamper, 113–46. Budapest: Central European University Press, 2014.

Jędrzejewska, Bogumiła and Tomasz Samojlik. 'Żubr – puszcz bogaty skarb'. In *Ochrona i łowy. Puszcza Białowieska w czasach królewskich*, edited by Tomasz Samojlik, 75–88. Białowieża: Zakład Badania Ssaków PAN, 2005.

Jordan, Marion. *Andrei Platonov*. Letchworth: Bradda Books, 1973.

Josephson, Paul R. *Red Atom: Russia's Nuclear Power Program from Stalin to Today*. Pittsburgh: University of Pittsburgh Press, 2005.

Josephson, Paul, Nicolai Dronin, Aleh Cherp, Ruben Mnatsakanian, Dmitry Efremenko and Vladislav Larin. *An Environmental History of Russia*. Cambridge: Cambridge University Press, 2013.

Josephson, Paul. 'Introduction. The Stalin Plan for the Transformation of Nature, and the East European Experience'. In *In the Name of Great Work. Stalin's Plan for the Transformation of Nature and its Impact in Eastern Europe*, edited by Doubravka Olšáková, 1–36. New York: Berghahn Books, 2016.

Kaczorowski, Paweł and Paweł Gajewski. 'Górnictwo węgla kamiennego w Polsce w okresie transformacji'. *Acta Universitatis Lodziensis. Folia Oeconomica* 219 (2008): 201–27.

Kahn, Andrew, Lipovetsky, Mark, Reyfman, Irina and Stephanie Sandler. *A History of Russian Literature*. Oxford: Oxford University Press, 2018.

Kania, Waldemar. 'Mord'. *Życie Literackie* no 7 (1989): 6.

Kapralski, Sławomir. 'Amnesia, Nostalgia, and Reconstruction: Shifting Modes of Memory in Poland's Jewish Spaces'. In *Jewish Space in Contemporary Poland*, edited by Erica T. Lehrer and Michael Meng, 149–69. Bloomington: Indiana University Press, 2015.

Kataev, Valentin. *Time, Forward!* New York: Farrar & Rinehart, 1933 [1932].

Kennedy, Rosanne. 'Multidirectional Eco-Memory in an Era of Extinction. Colonial Whaling and Indigenous dispossession in Kim Scott's "That Deadman Dance"'. In *The Routledge Companion to the Environmental Humanities*, edited by Ursula K. Heise, Jon Christensen and Michelle Niemann, 268–77. London: Routledge, 2017.

Klementowski, Robert. *W cieniu sudeckiego uranu. Kopalnictwo uranu w Polsce w latach 1948–1973*. Wrocław: Instytut Pamięci Narodowej, 2010.

Klier, John. 'The Holocaust and the Soviet Union'. In *The Historiography of the Holocaust*, edited by Dan Stone, 276–95. Basingstoke: Palgrave Macmillan, 2004.

Klingle, Matthew. 'The Nature of Desire. Consumption in Environmental History'. In *The Oxford Handbook of Environmental History*, edited by Andrew C. Isenberg, 467–512. Oxford: Oxford University Press, 2017.

Kobielska, Maria. 'Czytanie Nory. Appendix'. In *Od pamięci biodziedzicznej do postpamięci*, edited by Teresa Szostek, Roma Sendyka and Ryszard Nycz, 189–205. Warszawa: IBL PAN, 2013.

Koepnick, Lutz. *The Long Take: Art Cinema and the Wondrous*. Minneapolis: University of Minnesota Press, 2017.

Kolbuszewski, Jacek. *Krajobraz i kultura. Sudety w literaturze i kulturze polskiej*. Katowice: Wydawnictwo "Śląsk", 1985.

Konarski, Feliks. 'Katyń'. *Tydzień* no 19 (1998): 1.

Konczal, Agata Agnieszka. *Antropologia lasu. Leśnicy a percepcja i kształtowanie wizerunków przyrody w Polsce*. Warszawa: Wydawnictwo IBL PAN, 2017.

Konwicki, Tadeusz. *Kompleks Polski*. Warszawa: Wydawnictwo Alfa, 1990 [1977].

Kopanic, Michael. 'Industrialization'. In *Encyclopedia of Eastern Europe: From the Congress of Vienna to the Fall of Communism*, edited by Richard Frucht, 322–3. New York: Garland Publishing, 2000.

Kotkin, Stephen. *Magnetic Mountain: Stalinism as a Civilization*. Berkeley: University of California Press, 1995.

Kosterska, Alicja. 'Tekstowa hybryda jako medium pamięci. O „Miedziance. Historii znikania" Filipa Springera'. *Zeszyty Naukowe Towarzystwa Doktorantów UJ. Nauki Humanistyczne* 11 (2015): 7–26.

Kovacs, Andras Balint. *The Cinema of Bela Tarr. The Circle Closes*. London: Wallflower Press, 2013.

Krasznahorkai, László. *Satantango*. Translated by George Szirtes. London: Atlantic Books, 2012.

Krzysztofik, Robert and Mirek Dymitrow (eds). *Degraded and Restituted Towns in Poland: Origins, Development, Problems.* Gothenburg: University of Gothenburg, 2015.

Kula, Marcin. *Nośniki pamięci historycznej.* Warszawa: DiG, 2002.

Kühne, Olaf. 'Przestrzeń, krajobraz i krajobraz kulturowy. Terminologia, definicje'. In *Krajobrazy kulturowe. Sposoby konstruowania i narracji,* edited by Robert Traba, Violetta Julkowska and Tadeusz Styjakiewicz, 25–45. Warszawa-Berlin: Neriton, 2017.

Lane, Peter. *The USSR in the Twentieth Century.* London: Pavilion Books, 1978.

Lanzmann, Claude. 'Le Lieu et la parole'. In *Au sujet de Shoah: Le Film de Claude Lanzmann,* edited by Michel Deguy. Paris: Editions Berlin, 1990.

Latour, Bruno. *Facing Gaia. Eight Lectures on the New Climatic Regime.* Translated by Catherine Porter. Cambridge: Polity Press, 2017.

Lebda, Magdalena. *Granica lasu.* Poznań: Wydawnictwo Wojewódzkiej Biblioteki Publicznej, 2013.

LeBlanc, Ronald. 'The Mikoyan Mini-Hamburger, or How the Socialist Realist Novel about the Soviet Meat Industry Was Created'. *Gastronomica: The Journal of Critical Food Studies* 16, no. 2 (2016): 31–44.

Lee Klein, Kerwin. 'On the Emergence of Memory in Historical Discourse'. *Representations* 69, Special Issue: 'Grounds for Remembering' (2000): 127–50.

Lindbladh, Johanna (ed.). *The Poetics of Memory in Post-Totalitarian Narration.* Lund: The Centre for European Studies, 2008a.

Lindbladh, Johanna. 'The Problem of Narration and Reconciliation in Svetlana Aleksievich's Testimony *Voices from Chernobyl*'. In *The Poetics of Memory in Post-Totalitarian Narration,* edited by Johanna Lindbladh, 41–53. Lund: The Centre for European Studies, 2008b.

Małczyński, Jacek. 'Odwracanie krajobrazu, Rewitalizacja miejsc pamięci Holocaustu w sztuce współczesnej na przykładzie "Winterreise" Mirosława Bałki oraz Magdaleny Hueckel i Tomasza Śliwińskiego'. In *Od pamięci biodziedzicznej do postpamięci,* edited by Teresa Szostek, Roma Sendyka and Ryszard Nycz, 223–37. Warszawa: Instytut Badań Literackich PAN, 2013.

Małczyński, Jacek. 'Historia środowiskowa Zagłady'. *Teksty Drugie* 164, no. 2 (2017): 17–33.

Małczyński, Jacek, Ewa Domańska, Mikołaj Smykowski and Agnieszka Kłos. 'The Environmental History of the Holocaust'. *Journal of Genocide Research*. Published online: 22 January 2020. 1–14.

Mannan, Sam. *Lees' Process Safety Essentials. Hazard Identification, Assessment and Control.* Amsterdam: Elsevier, 2013.

Manser, Roger. *Failed Transitions: The Eastern European Economy and Environment Since the Fall of Communism.* New York: New Press, 1994.

Masing-Delic, Irene. *Abolishing Death. A Salvation Myth of Russian Twentieth-Century Literature.* Stanford: Stanford University Press, 1992.

Maslowski, Michel. 'Introduction: Invention de l'Europe Centrale'. In *Culture et identite en Europe centrale. Canon litteraires et visions de l'histoire*, edited by Michel Maslowski, Didier Francfort and Paul Gradvohl, 13-32. Paris: Institut d'études slaves, 2011.

Massumi, Brian. *Parables for the Virtual. Movement, Affect, Sensation*. Durham NC: Duke University Press, 2002.

McCarthy, Katherine. 'Environment'. In *Encyclopedia of Eastern Europe: From the Congress of Vienna to the Fall of Communism*, edited by Richard Frucht, 219-20. New York: Garland Publishing, 2000.

McKenzie, Wark. *Molecular Red: Theory for the Anthropocene*. London: Verso, 2015.

McNeill, J. R. and Corinna Unger (eds). *Environmental Histories of the Cold War*. New York: Cambridge University Press, 2010.

McNeill, J. R. and Erin Stuart Mauldin (eds). *A Companion to Global Environmental History*. Chichester: Blackwell Publishing, 2012.

McNeill, J. R. and Peter Engelke. *The Great Acceleration. An Environmental History of the Anthropocene since 1945*. Cambridge MA: The Belknap Press of Harvard University Press, 2014.

Medvedev, Zhores. *The Legacy of Chernobyl*. New York: W. W. Norton, 1992.

Mickiewicz, Adam. *Pan Tadeusz or the Last Foray in Lithuania. A Story of Life among Polish Gentlefolk in the Years 1811 and 1812 in Twelve Books*. Translated by George Rapall Noyes. London: J. M. Dent & Sons, 1917 [1834]. Available online: https://www.gutenberg.org/files/28240/28240-pdf.pdf

Miłosz, Czesław. 'Odbicia'. In *Wiersze*, vol. 1. Kraków: Wydawnictwo Literackie, 1987.

Młyńczak Halina. 'Katyńskie kwiaty'. In Helska Latarnia, edited by Jan Zdzisław Brudnicki and Robert Zapora, 97. Warszawa: COSM, 2002.

Mocanu, Sorin. 'The Masks of Collectivization. Dramatization of Income Distribution'. *Acta Iassyensia Comparationis* 9 (2011): 173-84.

Mońko, Michał. 'Gułag Miedzianka'. *Odra* 4 (1995): 33-9.

Moon, David. 'The Curious Case of the Marginalisation or Distortion of Russian and Soviet Environmental History in Global Environmental Histories'. *International Review of Environmental History* 3, no. 2 (2017): 31-50.

Moore, Irina. 'Vilnius Memoryscape: Razing and raising of monuments, collective memory and national identity'. *Linguistic Landscape* 5 no. 3 (2019): 248-80.

Morcinek, Gustaw. *Łysek z Pokładu Idy. Opowiadanie*. Lwów: Państwowe Wydawnictwo Książek Szkolnych, 1933.

Morton, Timothy. *Hyperobjects. Philosophy and Ecology after the End of the World*. Minneapolis: University of Minnesota Press, 2013.

Morton, Timothy. *Being Ecological*. Cambridge MA: MIT Press, 2018.

Motyka, Grzegorz. *Wołyń '43. Ludobójcza czystka – fakty, analogie, polityka historyczna*. Kraków: Wydawnictwo Literackie, 2016.

Mozhaev, Boris A. and Alexander Solzhenitsyn. *Lively and Other Stories and a Memoir by Alexander Solzhenitsyn*. Translated by David Holohan. Surbiton: Hodgson Press, 2008.

Muzaini, Hamzah and Brenda Yeoh. *Contested Memoryscapes. The Politics of Second World War and Commemoration in Singapore.* London: Routledge, 2016.

Mycio, Mary. *Wormwood Forest. A Natural History of Chernobyl.* Washington: Joseph Henry Press, 2005.

Naimark, Norman. 'The Sovietization of Eastern Europe, 1944–1953'. In *The Cambridge History of the Cold War. Vol. 1 Origins,* edited by Melvyn P. Leffler and Odd Arne Westad, 175–97. Cambridge: Cambridge University Press, 2010.

Neirick, Miriam. *When Pigs Could Fly and Bears Could Dance. A History of the Soviet Circus.* Madison: The University of Wisconsin Press, 2012.

Newerly, Igor. *Zostało z uczty Bogów.* Warszawa: PIW, 2018 [1986].

Niklasson, Mats, Ewa Zin, Tomasz Zielonka, Markus Feijen, Adolf F. Korczyk, Marcin Churski, Tomasz Samojlik, Bogumiła Jędrzejewska, Jerzy M. Gutowski and Bogdan Brzeziecki. 'A 350-year tree-ring fire record from Białowieza Primeval Forest, Poland: implications for Central European lowland fire history'. *Journal of Ecology* 98 (2010): 1319–29.

Nikulin, Dmitri (ed.). *Memory: A History.* Oxford: Oxford University Press, 2015.

Nixon, Rob. *Slow Violence and the Environmentalism of the Poor.* Cambridge MA: Harvard University Press, 2011.

Nora, Pierre. *Les lieux de mémoire. La République* (1 vol., 1984), *La Nation* (3 vol., 1986), *Les France* (3 vol., 1992). Paris: Gallimard.

Ostashevsky, Eugene. 'Selections from the "Triumph of Agriculture"'. *American Poetry Review* Jul/Aug 34 no. 4 (2005): 31–3.

Parthé, Kathleen F. *Russian Village Prose. The Radiant Past.* Princeton: Princeton University Press, 1992.

'Partisans' (anonymous song). Reprint from *Ogniwo* no 33. Paris (September 1952): 10–11.

Patey-Grabowska, Alicja. 'Impresje Katyńskie (Pamięci ojca)'. *Zorza* no 27 (1989): 20.

Pick, Anat and Guinevere Narraway. 'Introduction. Intersecting Ecology and Film'. In *Screening Nature. Cinema beyond the Human,* edited by Anat Pick and Guinevere Narraway, 1–18. New York: Berghahn Books, 2013.

Platonov, Andrey. 'The Wanderer'. Translated by Albert C. Todd. In *Twentieth Century Russian Poetry. Silver and Steel. An Anthology,* edited by Yevgeny Yevtushenko, Albert Todd and Max Hayward, 348. New York: Doubleday, 1993.

Platonov, Andrey. *The Foundation Pit.* Translated by Robert Chandler, Elizabeth Chandler and Olga Meerson. London: Vintage Books, 2010.

Plumwood, Val. 'The Cemetery Wars: Cemeteries, Biodiversity and the Sacred'. *Local-Global: Identity, Security, Community* 3 (2007): 54–71.

Pollack, Martin. *Kontaminierte Landschaften. Unruhe bewahren.* St Pölten and Wien: Residenz Verlag, 2014. (All references from Polish edition: *Skażone krajobrazy.* Translated by Karolina Niedenthal. Wołowiec: Wydawnictwo Czarne, 2014)

Połońska, Elżbieta. 'W pętli' (In the Noose). In *Antologia poezji radzieckiej,* vol. 1, edited by Józef Waczków, 119–23. Warszawa: PIW, 1979.

Pomian, Krzysztof. 'Historia – dziś'. In *Historia – dziś. Teoretyczne problemy wiedzy o przeszłości*, edited by Ewa Domańska, Rafał Stobiecki and Tomasz Wiślicz, 19–36. Kraków: Universitas, 2014.

Poolos, Jamie. *The Atomic Bombings of Hiroshima and Nagasaki*. New York: Chelsea House, 2008.

Pryde, Philip Rust. *Environmental Management in the Soviet Union*. Cambridge: Cambridge University Press, 1991.

Pyssa, Justyna. 'Uranium Occurrence, Deposits and Mines in Poland'. *Inżynieria Mineralna* 17, no 1 (2016): 47–55.

Raffnsøe, Sverre. *Philosophy of the Anthropocene. The Human Turn*. London: Palgrave Macmillan, 2016.

Rapson, Jessica. 'Fencing in and Weeding Out: Curating Nature at Former Concentration Camps in Europe'. In *Emerging Landscapes. Between Production and Representation*, edited by Davide Deriu, Krystallia Kamvasinou and Eugénie Shinkle, 161–71. Farnham: Ashgate, 2014.

Redliński, Edward. *Konopielka*. Warszawa: Ludowa Spółdzielnia Wydawnicza, 1973.

Rigby, Kate. 'Tragedy, Modernity, and "Terra Mater": Christa Wolf Recounts the Fall'. *New German Critique* 101, vol. 34, no 2 (2007): 115–41.

Rigby, Kate. *Dancing with Disaster: Environmental Histories, Narratives, and Ethics for Perilous Times*. Charlottesville: University of Virginia Press, 2015.

Roskin, Michael G. *The Rebirth of East Europe*. Upper Saddle River: Pearson Education, 2002.

Rothberg, Michael. *Traumatic Realism. The Demands of Holocaust Representation*. Minneapolis: University of Minnesota Press, 2000.

Rothberg, Michael. *Multidirectional Memory. Remembering the Holocaust in the Age of Decolonisation*. Stanford: Stanford University Press, 2009.

Różewicz, Tadeusz. 'Most płynie do Szczecina'. In *Wejście w kraj. Wybór reportaży z lat 1944–1964*, edited by Zbigniew Stolarek, vol. I, Warszawa: Iskry, 1965.

Różewicz, Tadeusz. *Niedzielny spacer za miasto*. In *Proza*. Wrocław: Zakł. Narodowy im. Ossolińskich, 1973.

Różycki, Tomasz. 'Scorched Maps'. Translated by Mira Rosenthal. In *Colonies*. Brookline: Zephyr Press, 2013.

Rymkiewicz, Jarosław Marek. *Rozmowy polskie latem roku 1983*. Warszawa: Bellona, 2009.

Samojlik, Tomasz, Ian Rotherham and Bogumiła Jędrzejewska. 'Quantifying Historic Human Impacts on Forest Environments: A Case Study in Bialowieza Forest, Poland'. *Environmental History* 18, no. 3 (2013): 576–602.

Schama, Simon. *Landscape and Memory*. New York: A. A. Knopf, 1995.

Schramm, Katharina. 'Landscapes of Violence: Memory and Sacred Space'. *History and Memory* 23, no. 1 (2011): 5–22.

Seifrid, Thomas. *Andrei Platonov: Uncertainties of Spirit*. Cambridge: Cambridge University Press, 1992.

Seifrid, Thomas. *A Companion to Andrei Platonov's The Foundation Pit*. Boston: Academic Studies Press, 2009.

Sendyka, Roma. 'Miejsca, które straszą (afekty i nie-miejsca pamięci)'. *Teksty Drugie* 143, no. 1 (2014): 84–102.

Sendyka, Roma. 'Krajobrazy (nie)pamięci: dekonstrukcja krajobrazu kulturowego'. In *Więcej niż obraz*, edited by Eugeniusz Wilk, Anna Nacher, Magdalena Zdrodowska, Ewelina Twardoch and Michał Gulik, 81–99. Gdańsk: Katedra, 2015a.

Sendyka, Roma. 'Prism: Understanding Non-Sites of Memory'. Translated by Jennifer Croft. *Teksty Drugie* 2 (2015b): 13–28.

Słoński, Edward. Na zgliszczach [On the Ashes]. In *Rozkwitały pąki białych róż ... Wiersze i pieśni lat 1908–1918 o Polsce, o wojnie i o żołnierzach*, edited by Andrzej Romanowski. Warszawa: Czytelnik, 1990.

Smykowski, Mikołaj. 'Eksterminacja przyrody w Lesie Rzuchowskim'. *Teksty Drugie* 164, no. 2 (2017): 61–85.

Snyder, Timothy. *Bloodlands. Europe between Hitler and Stalin*. New York: Basic Books, 2010.

Soloukhin, Vladimir. 'To Make Birds Sing'. Translated by Daniel Weissbort. In *Twentieth Century Russian Poetry. Silver and Steel. An Anthology*, edited by Yevgeny Yevtushenko, Albert Todd and Max Hayward, 753. New York: Doubleday, 1993.

Soroczyński, Tadeusz. 'Zapis Puszczy'. In *XX Jubileuszowy Kalendarz Opolski*. Opole: Otko, 1978.

Sörlin, Sverker and Paul Warde. *Nature's End. History and the Environment*. New York: Palgrave Macmillan, 2009.

Springer, Filip. *History of a Disappearance: The Story of a Forgotten Polish Town*. Translated by Sean Bye. New York: Restless Books, 2017.

Stalin, Joseph Vissarionovich. *The essential Stalin. Major Theoretical Writings 1905–1952*, edited by Bruce Franklin. London: Croom Helm, 1973.

Stasiuk, Andrzej. *Wschód*. Wołowiec: Czarne, 2014.

Stoetzer, Bettina. 'A Path through the Woods: Remediating Affective Landscapes in Documentary Asylum Worlds'. *Transit: A Journal of Travel, Migration and Multiculturalism in the German-speaking World* 9, no. 2 (2014): 1–23.

Stoetzer, Bettina. 'Ruderal Ecologies: Rethinking Nature, Migration, and the Urban Landscape in Berlin'. *Cultural Anthropology* 33, no. 2 (2018): 295–323.

Sturdy, Caroline. *Holocaust Archeologies. Approaches and Future Directions*. New York: Springer, 2015.

Szczepan, Aleksandra. 'Krajobrazy postpamięci'. *Teksty Drugie* 143, no. 1 (2014): 103–26.

Szpak, Ewelina. *Między osiedlem a zagrodą. Życie codzienne mieszkańców PGR-ów*. Warszawa: Trio, 2005.

Szwagrzyk, Jerzy. 'Bialowieza Forest: what it used to be, what it is now and what we want it to be in the future'. *Forest Research Papers* 77, no. 4 (2016): 291–5.

Tokarczuk, Olga. *Drive Your Plow Over the Bones of the Dead*. Translated by Antonia Lloyd-Jones. London: Fitzcarraldo, 2019.

Tomiałojć, Ludwik and Tomasz Wesołowski. 'Diversity of the Bialowieza forest avifauna in space and time'. *Journal of Ornithology* 145 (2004): 81–92.

Topolski, Jerzy. 'Problemy metodologiczne korzystania ze źródeł literackich w badaniu historycznym'. In *Dzieło literackie jako źródło historyczne*, edited by Zofia Stefanowska and Janusz Sławiński, 8–23. Warszawa: Czytelnik, 1978.

Traba, Robert. 'Kulturlandschaften erinnern – von der Aneignung über das "Depositum" bis zur "geistigen Nachfolge/aktiven Erbschaft"'. In *Cultural Landscapes. Transatlantische Perspektiven auf Wirkungen und Auswirkungen deutscher Kultur und Geschichte im östlichen Europa*, edited by Andrew Demshuk and Tobias Weger, 29–39. München: De Gruyter, 2015.

Traba, Robert and H. H. Hahn (eds). *Deutsch-Polnische Erinnerungsorte*. Paderborn: Schöningh Paderborn, 2015–17.

Traba, Robert. 'Krajobraz kulturowy: strategie badawcze i interpretacje'. In *Krajobrazy kulturowe. Sposoby konstruowania i narracji*, edited by Robert Traba, Violetta Julkowska and Tadeusz Styjakiewicz, 11–24. Warszawa-Berlin: Neriton, 2017.

Trexler, Adam. *Anthropocene Fictions: The Novel in a Time of Climate Change*. Charlottesville: University of Virginia Press, 2015.

Tribe, Keith. *The Economy of the Word. Language, History, and Economics*. Oxford: Oxford University Press, 2015.

Trojanowska, Tamara, Joanna Niżyńska and Przemysław Czapliński (eds). *Being Poland. A New History of Polish Literature since 1918*. Toronto: University of Toronto Press, 2018.

Tuan, Yi-Fu. *Landscapes of Fear*. Minneapolis: University of Minnesota Press, 2013 [1979].

Tuan, Yi-Fu. *Space and Place. The Perspective of Experience*. Minneapolis: University of Minnesota Press, 2014 [1977].

Tumarkin, Maria. *Traumascapes: The Power and Fate of Places Transformed by Tragedy*. Victoria: Melbourne University Publishing, 2015.

Twardoch, Szczepan. *Drach*. Kraków: Wydawnictwo Literackie, 2014.

Ubertowska, Aleksandra. '"Kamienie niepokoją się i stają się agresywne". Holokaust w świetle ekokrytyki'. *Poznańskie Studia Polonistyczne. Seria Literacka* 45, no. 25 (2015): 93–111.

van Alphen, E. *Armando: Shaping Memory*. Rotterdam: NAi Publishers, 2000.

Veyne, Paul. *Writing History. Essay on Epistemology*. Translated by Mina Moore-Rinvolucri. Middletown: Wesleyan University Press, 1984 [1971].

Vitale, Serena. *Shklovsky: Witness to an Era*. Translated by Jamie Richards. Champaign: Dalkey Archive Press, 2012.

Voznesensky, Andrei. *On the Edge. Poems and Essays from Russia*. Translated by Richard McKane. London: Weidenfeld & Nicolson, 1991.

Wallace, Molly. *Risk Criticism: Precautionary Reading in an Age of Environmental Uncertainty*. Ann Arbor: University of Michigan Press, 2016.

Wantuła, Leon. *Urodzeni w dymach*. Katowice: Wydawnictwo "Śląsk", 1965.

Waterton, Emma. 'Landscape and Non-representational Theories'. In *The Routledge Companion to Landscape Studies*, edited by Peter Howard, Ian Thompson and Emma Waterton, 66–75. London: Routledge, 2013.

Webster, Barbara Jancar. *Environmental Action in Eastern Europe: Responses to Crisis*. London: Routledge, 2016.

Weiner, Douglas. *Models of Nature. Ecology, Conservation and Cultural Revolution in Soviet Russia*. Pittsburgh: University of Pittsburgh Press, 2000.

Wesołowski, T., Czeszczewik, D., Mitrus, C., Rowinski, P. 'Birds of the Białowieza National Park'. *Notatki Ornitologiczne* 44 (2003): 1–31.

White, Hayden. *Metahistory. The Historical Imagination in Nineteenth-Century Europe*. Baltimore: The John Hopkins University Press, 1973.

White, Hayden. *Tropics of Discourse*. Baltimore: The John Hopkins University Press, 1986.

White, Hayden. *The Practical Past*. Evanston: The Northwestern University Press, 2014.

Więcko, Edward. *Puszcza Białowieska*. Warszawa: Państwowe Wydawnictwo Naukowe, 1984.

Wirth, Peter, Barbara Mali and Wolfgang Fischer. *Post-Mining Regions in Central Europe – Problems, Potentials, Possibilities*. München: Oekom Verlag, 2012.

Wittlin, Józef. *Salt of the Earth*. Translated by Patrick Corness. London: Pushkin Press, 2019 [1935].

Wolf, Christa. *Accident. A Day's News*. Translated by Heike Schwarzbauer and Rick Takvorian. Chicago: The University of Chicago Press, 2001.

Wylegała, Anna. 'O perspektywach badania chłopskiego doświadczenia reformy rolnej. Z warsztatu badawczego'. *Rocznik Antropologii Historii* VII, no. 10 (2017): 273–305.

Yevtushenko, Yevgeni. 'Babiy Yar'. Translated by Robin Milner-Gulland and Peter Levi. In *Selected Poems*, 82–84. Harmondsworth: Penguin Books, 1962.

Young, Georges M. *The Russian Cosmists. The Esoteric Futurism of Nikolai Fedorov and His Followers*. Oxford: Oxford University Press, 2012.

Yuzefpolskaya, Sofija and George Rueckert. 'No Empty Game: The Immortality of the Poet in Arseny Tarkovsky's Memorial Poems to N. A. Zabolotsky and A. A. Akhmatova'. *The Slavic and East European Journal* 50, no. 2 (2006): 287–308.

Zabolotsky, Nikolay. 'Metamorphoses'. Translated by Daniel Weissbort. *Poetry* (July 1974).

Zabolotsky, Nikolai. 'The Face of the Horse' [1926]. Translated by Daniel Weissbort. In *Twentieth Century Russian Poetry. Silver and Steel. An Anthology*, edited by Yevgeny Yevtushenko, Albert Todd and Max Hayward, 450–1. New York: Doubleday, 1993.

Zabolotsky, Nikolay. *Selected Poems*. Translated by Daniel Weissbort, Robin Milner-Gulland and Peter Levi. Manchester: Carcanet Press, 1999.

Zabuzhko, Oksana. 'Planeta Piołun – Dowżenko – Tarkowski – Von Trier albo dyskurs nowej grozy'. Translated [into Polish] by Katarzyna Kotyńska. In *Po Czarnobylu*.

Miejsce katastrofy w dyskursie współczesnej humanistyki, edited by Iwona Boruszkowska, Katarzyna Glinianowicz, Aleksandra Grzemska and Paweł Krupa, 36–51. Kraków: Wydawnictwo Uniwersytetu Jagiellońskiego, 2017.

Zapf, Hubert. 'Literature as Cultural Ecology: Notes towards a Functional Theory of Imaginative Texts with Examples from American Literature'. *REAL* 17 (2001): 85–100.

Ziegler, Charles. *Environmental Policy in the USSR*. Amherst: University of Massachusetts Press, 1987.

Zink, Andrea. 'Approaching the Void – Chernobyl' in Text and Image'. *Anthropology of East Europe Review: Memories, commemorations, and representations of Chernobyl* 30, no. 1 (2012): 100–12.

Zub, Karol. 'Ssaki'. In *Białowieski Park Narodowy. Poznać – Zrozumieć – Zachować*. Białowieża: Białowieski Park Narodowy, 2009.

Miscellaneous Items

All available online materials last accessed 18 May 2020

Alexievich, Svetlana. *Biographical*. The Nobel Prize in Literature 2015. Available online: https://www.nobelprize.org/prizes/literature/2015/alexievich/biographical/

Appunn, Kerstine. 'Germany's Three Lignite Mining Regions'. Clear Energy Wire, 2018. https://www.cleanenergywire.org/factsheets/germanys-three-lignite-mining-regions

Aseh.net. The website of American Society for Environmental History. Available online: https://aseh.net/about-aseh/mission-statement

Asle.org. The website of Association for the Study of Literature and Environment. Available online: https://www.asle.org/discover-asle/a-message-from-the-asle-president/

Belovezhskaya pushcha by Pesniary. Composed by Aleksandra Pakhmutova. Lyrics: Nikolai Dobronravov, 1975. Translated by Mikhail Palstsianau. lyrictranslate.com. Available online: https://lyricstranslate.com/en/%D0%B1%D0%B5%D0%BB%D0%BE%D0%B2%D0%B5%D0%B6%D1%81%D0%BA%D0%B0%D1%8F-%D0%BF%D1%83%D1%89%D0%B0-belovezhskaza-pushcha-belovezhskaya-pushcha.html

Białowieża Forest. UNESCO. Available online: https://whc.unesco.org/en/list/33/documents/

Bock-Schroeder, Peter. *Disturbed Landscapes*. Bock-Schroeder.Com 1964. Available online: http://bock-schroeder.com/disturbed-landscapes

Chernobyl Nuclear – "Surviving Disaster". Nick Murphy (Dir.), BBC Drama / Documentary, 2006.

EASLCE.eu. The website of European Association for Studies of Literature, Culture and the Environment. Available online: https://www.easlce.eu/about-us/our-aims/

Environmental History journal website: https://academic.oup.com/envhis/pages/About

Eseh.org. The website of European Society for Environmental History. Available online: http://eseh.org/about-eseh/mission/

Horoszkiewicz, Janusz. *Szlakiem wołyńskich krzyży – Ugły.* Isakowicz.pl 2016. Available online: http://isakowicz.pl/szlakiem-wolynskich-krzyzy-ugly/

Hunting. Polish State Forests 2016. Available online: https://www.lasy.gov.pl/en/our-work/hunting

IPN. Upamiętnianie walk i męczeństwa. IPN.GOV.PL 2016. Available online: https://ipn.gov.pl/pl/upamietnianie/38749,Upamietnianie-walk-i-meczenstwa.html

Łuskino, Leon. 'Szara piechota' (a song). A-pesni.org. Available online: http://a-pesni.org/polsk/maszeruja.htm [1918]

Makarski, Antoni. 'O żołnierzu' (a song). Polish National Armed Forces, 2010. Available online: https://www.nsz.com.pl/index.php/biblioteka/146-piosenki-i-wiersze-nsz

McLellan, B. N., Proctor, M. F., Huber, D. and Michel, S. *Ursus arctos* (amended version of 2017 assessment). *The IUCN Red List of Threatened Species* 2017: e.T41688A121229971. Available online: http://dx.doi.org/10.2305/IUCN.UK.2017-3.RLTS.T41688A121229971.en

Miedzianka: Stalin's revenge for uranium (a radio play), 2008. Available online: https://www.radioram.pl/articles/view/6720/Miedzianka-zemsta-Stalina-za-uran-Sluchowisko

National Park 'Belovezhskaya Pushcha'. *History.* Available online: https://npbp.by/eng/about/history/

Norman Davies on Polish History. Interview. Available online: https://www.youtube.com/watch?v=3OoSdnebLxw

Oswiecim.pl. *Oświęcimska przestrzeń spotkań – projekty.* Available online: https://oswiecim.pl/ops-projekty/

Polish Central Statistical Office, 2017. *Production of Brown Coal.* Available online: http://stat.gov.pl/obszary-tematyczne/przemysl-budownictwo-srodki-trwale/przemysl/produkcja-wyrobow-przemyslowych-w-2015-roku,3,13.html *Production of Hard Coal.* Available online: https://stat.gov.pl/obszary-tematyczne/przemysl-budownictwo-srodki-trwale/przemysl/produkcja-wazniejszych-wyrobow-przemyslowych-w-kwietniu-2017-roku,2,61.html

Ritchie, Hannah and Max Roser. 'Meat and Seafood Production & Consumption'. OurWorldInData.org, 2019. Available online: https://ourworldindata.org/meat-and-seafood-production-consumption#empirical-view

René, Miloš. 'History of Uranium Mining in Central Europe, Uranium - Safety, Resources, Separation and Thermodynamic Calculation'. *IntechOpen*, 2017. DOI: 10.5772/intechopen.71962. Available online: https://www.intechopen.com/books/uranium-safety-resources-separation-and-thermodynamic-calculation/history-of-uranium-mining-in-central-europe

Rodak, Anna. 'Las'. Prawicowyinternet.pl 2015. Available online: https://prawicowy internet.pl/poezja-katynska/

Rozovsky, Liza. 'How to Commemorate the Nazi's Largest-ever Jewish Massacre?' Haaretz.com. Published Jan 30, 2020. Available online: https://www.haaretz.com/israel-news/.premium.MAGAZINE-ukrainian-center-to-commemorate-babi-yar-largest-ever-jewish-massacre-1.8448655

Sátántangó (Satantango). Béla Tarr (Dir.), László Krasznahorkai (script), Vega Film 1994.

Sirota, Lyubov. 'The Chernobyl Poems'. Translated by Leonid Levin and Elisavietta Ritchie. 2003. Available online: https://brians.wsu.edu/2016/12/05/chernobyl-poems/

Springer, Filip. Radio interview. Ninateka.pl 2012. Available online: https://ninateka.pl/audio/filip-springer-swiat-mlodych

Son of Saul. László Nemes and Clara Royer (screenwriters), Hungarian National Film Fund, Laokoon Filmgroup 2015.

Stalker. Andrei Tarkovsky (Dir.), Mosfilm 1979.

Stasiuk, Andrzej. 'East, or, the Veins of this Land'. Translated by Irena Maryniak. *Eurozine* 1 March 2016. Available online: https://www.eurozine.com/east-or-the-veins-of-this-land/

The Battle of Chernobyl. Thomas Johnson (Dir.), Discovery Channel, 2006.

The Local.de. 'Germany backs speedier "shutdown plan" for coal mines'. 2020. Available online: https://www.thelocal.de/20200116/germany-looks-to-step-up-coal-exit-timetable

Tsvirko, Lidia. *Fauna Ecology*. Polesye State Radiation-Ecological Reserve. Available online: http://www.zapovednik.by/en/research/fauna/

Wołk-Jezierska, Witomiła. 'Garstka ziemi'. 2013. Available online: http://tl.bialowieza.pl/artykul/spotkanie-poswiecone-zbrodni-katynskiej

Żywiec.info. Powodzie na Żywiecczyźnie, 2014. Available online: http://zywiecinfo.pl/historia/item/1971-powodzie-na-zywiecczyznie

Newsreels

Spring Ploughing. 12/1946. Available online: http://www.repozytorium.fn.org.pl/?q=pl/node/4316

Colorado Beetle. 29/1948. Available online: http://www.repozytorium.fn.org.pl/?q=pl/node/5857

Breeding Economy. 7/1949. Available online: http://www.repozytorium.fn.org.pl/?q=pl/node/4982

The USSR. Cows' farm. 26/1949. Available online: http://www.repozytorium.fn.org.pl/?q=pl/node/5660

Steppes and Deserts into Farmlands. 41/1949. Available online: http://www.repozytorium.fn.org.pl/?q=pl/node/5511

Days of the Grain. 9/1951. Available online: http://www.repozytorium.fn.org.pl/?q=pl/node/7185

Fight for Fertility. 21/1951. Available online: http://www.repozytorium.fn.org.pl/?q=pl/node/7086

Harvest of Peace. 30/1951. Available online: http://www.repozytorium.fn.org.pl/?q=pl/node/7123

Mechanized Haymaking. 38/1951. Available online: http://www.repozytorium.fn.org.pl/?q=pl/node/6839

The USSR. Sowing for Peace. 21/1952. Available online: http://www.repozytorium.fn.org.pl/?q=pl/node/7385

The USSR. On the Fields of Kazakhstan. 44/1952. Available online: http://www.repozytorium.fn.org.pl/?q=pl/node/7679

5 marca 1953 roku zmarł Józef Stalin. 11–12/1953. Available online: http://www.repozytorium.fn.org.pl/?q=pl/node/7434

Index

Abramov, Fedor 73
Agamben, Giorgio 72
agriculture 20, 23, 35, 39, 73
   collectivized 36, 65, 67, 73, 42, 54, 59–60, 72
   socialist 59–63, 65–6
   see also collectivisation
Alaimo, Stacy 101
Alexievich, Svetlana 9, 136
   Chernobyl Prayer: A Chronicle of the Future 5, 136, 147–58
Andruchovych, Juriji 10
Angus, Jan 101
animals
   animal abuse 64, 79, 87–8
   bees 147–9, 173–4
   birds 46, 48, 73, 79, 109, 148, 154, 156–8, 181–3, 194
   brown bear 194, 200–1
   collectivization 4, 25, 37, 49–55, 62–6, 71, 85–6, 89–90
   cows 4, 52, 65, 71–2, 85
   emancipation 62–3, 87, 89
   European bison 194, 196, 198, 200–1
   flies 88
   Grey wolves 131, 194, 198, 200
   hamster 157
   history 12, 23, 51, 84
   horses 46, 49–54, 64–5, 72, 79, 89, 102, 116, 131, 156, 194
   killing 71, 130, 156
   lynx 194, 200
   memory 28, 156
   naming 70
   owl 89
   and peasants 36–7, 57, 67–8, 73, 76–8
   pigs 108–9
   protection 196, 201
   rabbits 119
   roes 174
   spiders 88
   translating 147–8

worms 148
   see also Białowieża Forest, Chernobyl, rural culture
Ankersmit, Frank 16–17, 27
Annales School 21
Anthropocene 40, 45, 101–2, 133–4, 136
   anthropocenic narrative 98, 107, 113–14, 122–3
   see also Silesia, Stalinocene
anthropocentric historiography 11–12, 17, 23, 25, 28, 84, 179, 187
anti-Semitism 165
Antonioni, Michelangelo 85
Applebaum, Anne
   Gulag 96
Appunn, Kerstine 93
Arctic Circle 100
ASEH (American Society for Environmental History) 22
Ash, Timothy Garton 10
ASLE (Association for the Study of Literature and Environment) 22
Assmann, Aleida 26, 170
Assmann, Jan 26–8, 170
atomic archipelago 103
Atommash 128
Auschwitz-Birkenau 152, 163, 177, 184–5

Babiy Yar 5, 165–7
   see also disturbed landscapes
Baer, Ulrich 169, 171–2, 175, 180
Bajcer, Stanisław 168
Bakhtin, Mikhail 148
   see also polyphony
Balassa, Bela A. 37
Bałka, Mirosław 174, 177, 185
Baratay, Éric 12
Barcz, Anna 98, 123
Barradas, Nuno 85
Bartoszyński, Kazimierz 15–16
Batori, Anna 80, 85–6, 88
Baudrillard, Jean 174

Bauman, Zygmunt 2
Baumgardten, Aleksander 161
Beck, Ulrich 94, 138, 151
Belarus 37, 128, 151, 153, 161, 167, 179, 185, 190, 192–3, 195–6, 200, 206
  see also Białowieża Forest, Chernobyl
Belavezha Accords 192
  see also Białowieża Forest, Viskuli
Bell, Peter D. 25
Belomorkanal see White Sea Canal
Belov, Andrei 73
Belov, Fedor 53
Belovezhskaya Pushcha see Białowieża Forest
Beloyarsk 128
Belyaev, Sergey 36
Bergthaller, Hannes 23
Białowieża Forest
  animals 190, 194
  conservation 190–1, 195, 197–8
  cultural landscape 5, 191–2, 200–3
  cultural memory 3, 191–2, 206
  Eastern European landscape 200, 207
  etymology 198–9
  foresters 195, 197–8
  history 190–3, 198, 203, 205–6
  Jadzwiez tribe 199
  literature 200–5
  National Park of Belovezhskaya Pushcha 190
  primeval forest 194–5, 197
  as World Heritage 195–6
  see also animals, Belarus, forest, Lithuania, pastoral, Poland, puszcza, trees
black cities 99
black triangle 96
Blackbourn, David 19, 186
Blacker, Uilleam 179
Blavascunas, Eunice 191, 197–8, 206
Bloch, Marc 11
Bohemia 96
Bohr, Niels 104
borderlands 94, 118, 174, 193, 199
Borzęcki, Robert 104
Brain, Stephen 12, 20–1, 127, 189
Bramwell, Anna 21
Breyfogle, Nicholas 22–3, 35
Brock, Emily K. 12

Brodsky, Joseph 68
Browarny, Wojciech 99
Brown, Deming 62, 68–9
Brown, Kate 127–8, 130, 143, 149
Bruno, Andy 19–23, 41, 130
  arctic environmental history 23
Buchta-Bartodziej, Petra 98, 123
Buell, Lawrence 28, 145, 183, 196
Bykivnia 162, 179

capitalism 19, 21, 40–1, 44, 97
Capitalocene 44, 101
  see also Anthropocene, Era of Man
Carey, Mark 12
Carson, Rachel
  Silent Spring 22
Carter, Francis W. 10
Caspian Sea 100
Catholic Church 37, 197
censorship 20, 61, 74, 120, 123, 128, 139, 156, 180, 183
Chakrabarty, Dipesh
  The Climate of History 11–12, 123
Charlesworth, Andrew 177
Cheliabinsk see Chelyabinsk
Cheloukhina, Svetlana 62–3
Chelyabinsk 25, 127–8, 130, 133
  see also East Ural Nature Reserve
Chełmoński, Józef 77
Chernobyl
  animals 133, 147–8, 156–7
  catastrophism 133–6, 152
  cloud 142, 144, 151, 153, 158
  cultural memory 127–38, 140–2, 145, 147, 150–1, 156, 158
  decontamination 130, 151, 157
  Eastern European risk narrative 127, 133, 143, 158
  ecological trauma 127, 131, 135, 139, 145, 147
  environmental history 133, 158
  environmental risk 127–8, 133–4, 136, 145
  heroism 129, 150
  literature 129–30, 133–8, 140, 144
  Madonna 151
  non-human world 131–3, 136, 156–8
  nuclear risk 128–30, 132, 134, 137, 141, 150

people 129, 151–2, 155, 157
poets 129, 134, 144
radiation 130–2, 134–5, 137, 139–40, 142, 144, 149–51, 158
reactor no. 4 129–30, 139, 143
representation 133–6, 138, 151
sarcophagus 130
sublimation 153
witnessing 127, 133, 135, 137–8, 140, 142–3, 147–58
see also Belarus, hyperobjects, contaminated language, Pripyat, Ukraine
Chukhontsev, Oleg 67–8
Clark, Katerina 59, 73–5
Clark, Timothy 102, 114, 134
climate 12, 23, 28, 35, 44, 60, 83, 93–4, 102
coal
   etymology 98–9
   extraction 4, 96, 98, 110–11
   history 96–7, 102
   narrative 94, 101, 110, 114, 122–3
   treasure 99, 112–13
   see also black triangle, fossil fuels, mining, Silesia
Cohen, Jeffrey Jerome 81, 134
Cold War 1–2, 19, 23, 27, 35, 61, 96, 103, 136–7, 150
Cole, Tim 178, 185
collectivization
   animals *see* animals' collectivization
   Belarus 37
   combinates 42
   commune 37
   cultural memory 37, 39, 54–7, 68, 71–2, 80
   Czechoslovakia 42, 90
   human-nature relationship 37, 39, 41–2, 45–56, 78, 90
   Hungary 37, 79, 90
   kolhoz 37, 42, 49, 51–3, 59–60, 64, 73
   Poland 37, 42, 90
   resistance 37, 57–8, 73, 78
   Soviet Union 25, 35, 37
   Sovkhoz 60
   toz 37
   Ukraine 36–7, 42, 90
   see also agriculture, Five-Year Plan, industrialization, contaminated language, peasantry, propaganda, rural culture, Stalinism, tired village
Collingwood, R. G. 11, 27
communism 4, 21, 24, 36, 51, 55, 57, 78–9, 93
   crisis 81–2, 86–7, 89–90
   post-communism 10, 61, 76, 94
   see also socialism, utopia
Conrad, Joseph
   Heart of Darkness 145
conservation movement
   as political opposition 20–1, 40
contaminated language 52, 55, 57, 59–61, 73, 76, 90, 142
   see also Chernobyl, propaganda
copper 103, 115–16, 118, 120
COP24 94
cultural landscape 41, 163, 170–1
'cursed soldiers' 168–9, 191
Czapliński, Przemysław 10, 15, 172, 184, 187
Czyżak, Agnieszka 111
Çağlayan, Emre 85–6

Danta, Darrick 80
Davies, Norman 199–200
deep-time 79, 83–4, 109–10
   see also geology, geological, natural or geological history
De Luca, Tiago 85
disturbed landscapes
   affective 164–5, 171, 174, 177
   aura 163, 172, 177, 180
   contaminated 164, 169, 171, 185
   cultural memory 162, 166–7, 188
   environmental history 169, 177, 185, 187
   environmental memory 164, 170, 180, 185–6, 188, 192–3
   memoryscapes 163, 165
   traumascapes 163, 165
   of violence 163–5, 177
   see also Babiy Yar, Katyń Forest, Volhynia
Długosz, Jan 199
Dobronravov, Nikolai 206
Domańska, Ewa 11, 173, 186
dragon 107–8, 112, 114
   see also Silesia's mythology
Duckert, Lowell 81, 98, 113, 123, 134
Dymitrow, Mirek 115

EASLCE (European Association for Studies of Literature, Culture and the Environment) 22
Easternness 10
East Ural Nature Reserve 132
　see also Chelyabinsk
Eastern Bloc see Soviet Bloc
Eastern European landscapes
　cultural memory 162–3, 167, 170, 172, 188
　ecocriticism 163, 167, 177–8, 193
　as environments 12, 162, 186–8
　see also borderlands, disturbed landscapes
ecocide 21, 131, 157
ecocritical perspective 24, 28, 97–8
ecocriticism 22–3, 27–9, 94
ecological
　memory see environmental memory
　scar (also trauma and wound) 13, 40, 107, 123, 182
　　see also Chernobyl, collectivization
　world 84, 86
　　see also non-human world
elements 75, 81, 83, 101, 104, 107–8, 122, 142, 158
Engelke, Peter 2, 19, 23, 40, 96, 103, 131, 134
Engels, Friedrich 41
environmental
　colonization 23, 72, 112, 192, 206
　cultures 23–6, 28, 31, 145, 192, 206–7
　　see also peasantry
　history 19, 21–2, 24–5, 30, 90, 191
　　see also Chernobyl, disturbed landscapes, history of non-humans, Silesia
　humanities 22–3, 101, 127
　imagination 28, 98
　memory 5, 24–5, 28, 31, 72, 76, 102, 105, 108, 170
　　see also disturbed landscapes
　risk see Chernobyl, mining, uranium
environmentalism 21, 197–8
Era of Man 101
　see also Anthropocene, Capitalocene, Great Acceleration
Errl, Astrid 26–8, 170
ESEH (European Society for Environmental History) 22

exhumations 167, 169, 179, 186
　see also disturbed landscapes
famine 25, 39, 47, 56, 67, 189
　in Ukraine, the *Holodomor* (Great Famine) 37, 53–4
Fenyo, Mario D. 79
Feshbach, Murray 10, 21, 59, 82, 128, 131
fiction 15–17, 31, 113–14, 136–7
Fischer, Wolfgang 115
Fitzpatrick, Sheila 37, 53–4, 58, 67, 94
Fiut, Aleksander 173
Five-Year Plan 25, 37, 39–40, 42, 50, 54–5, 58, 95
　The Great Rupture 39
　see also collectivization
Fleming, Michael 195
flowers 64, 72, 153, 168–9
forced labour camp see gulag
forest 21, 25, 74, 154, 169
　mourning with 180–4
　protection 189, 205–6
　shelter 178, 205
　see also Białowieża Forest, Chernobyl, Katyń Forest, *puszcza*, trees
fossil fuels 23, 35, 93, 102, 111, 143
　see also coal
Friendly, Alfred 10, 21, 59, 82, 128, 131
Frydryczak, Barbara 170
Fukushima 127
Fyodorov, Nikolai 64, 70

Garrard, Greg 40
geology 12, 21, 45, 79, 98–9, 101, 104, 113, 120, 124
　poetics (or narrative) 83–4, 102, 107–10, 114
　trauma (also wound) 97, 123
　victims 98
　voice 107–10, 112, 115
　see also Anthropocene, deep-time, natural or geological history
Germany
　East Germany 10, 96–7, 103, 144, 161
　Germans see Polonization, Silesia
　Nazi 40, 161–2, 167, 172–3, 175, 177, 179, 186, 203
Giddens, Anthony 94
Gifford, Terry 74, 183

Głowiński, Michał 16
Goldstein, Darra 63–4, 69
Gorbachev, Mikhail 128–9
Gorky, Maksim 40
Great Acceleration 2, 23, 39, 101, 134
    see also Anthropocene, Stalinocene
Great Hungarian Plain 80–1, 83, 85–6, 89
Greg, Wioletta 61
Grewe, Berndt-Stefan 12
Grochowiak, Stanisław
    Antiphon 204–5
Grossman, Joan Delaney 64
Grottger, Artur 203
gulag 44, 61, 94, 96–7, 120, 187
Györffy, M. 86

Halbwachs, Maurice 26–7
Handwerk, Gary 40
Hansen, Oskar 185
Hartwig, Julia 4, 71
    It will Speak 72
    Portrait I, II, III 71–2
heavy modernity 2
Heise, Ursula 94, 127, 138
Heller, Mikhail 36, 39, 53–4
heroism 25, 99
    see also Chernobyl
Hiroshima and Nagasaki 104, 150, 153
History
    definition 11–12
    discourse (and/or narrative) 11–12, 15–19, 23–8, 31
    and literature 15–18, 23–5
    natural or geological 12, 21, 82–4
        see also geology
    non-humans 11–12, 17, 22–4, 84, 90
    violence 162, 171, 183, 186, 192
    see also environmental history, fiction
Horoszkiewicz, Janusz 178–9
Hosking, Geoffrey 40, 73
Hueckel, Magda 185
Huk, Bogdan 178
Hundorova, Tamara 129, 152–4
    see also Chernobyl catastrophism
hydroengineering 40, 43, 45, 75–6, 100, 184
    see also rivers, White Sea Canal
hyperindustrialization 45–6, 60, 189
    see also collectivization, Stalinocene

hyperobjects 2, 133–5, 137, 140–2, 154–5
    narrative 134–5, 137, 145, 158
    see also Chernobyl
Hölderlin, Friedrich
    Remembrance 134

Illyés, Gyula 86
industrialization 9, 40–1, 44, 52, 69, 74, 77, 90, 97, 115
    see also agriculture, collectivization
Institute of National Remembrance (Poland) 167–9
iron 2, 94
Iron Curtain 3, 10, 123
    see also Cold War
Isenberg, Andrew C. 19, 22

Jaffe, Ira 85
Jaros, Jerzy 96
Jarosz, Dariusz 37
Jedwabne 162
Jędrzejewska, Bogumiła 190
Johnson, Thomas
    The Battle of Chernobyl 128
Jordan, Marion 44, 54
Josephson, Paul 9–10, 20–1, 36, 40–1, 44, 53–4, 67, 71, 128, 189

Kahn, Andrew 42, 61, 73
Kania, Waldemar
    Mord (Murder) 182
Kapralski, Sławomir 163
Kataev, Valentin
    Time, Forward! 95
Katyń Forest (also Katyń Massacre)
    cultural memory 16, 162, 179, 183
    executions 179
    poetry 180–4, 186–7
    see also disturbed landscapes, forest, trees
Kennedy, Rosanne 26–8
Kharkiv 179
Khrushchev, Nikita 21
Kiev 147, 164–5
    see also Babiy Yar
Klementowski, Robert 104
Kletno see Miedzianka, uranium
Klier, John 166
Klingle, Matthew 44

Kobielska, Maria 170
Koepnick, Lutz 85–6
Kolbuszewski, Jacek 161
Kolyma 97, 151
   see also gulag
Konarski, Feliks
   Katyń 181–2
Konczal, Agata Agnieszka 191, 197
Konwicki, Tadeusz
   Kompleks Polski (Polish Complex) 203–4
Kopanic, Michael 96
Kosterska, Alicja 117
Kotkin, Stephen 95
Kowary *see* Miedzianka, uranium
Kovacs, Andras Balint 85, 87–8
Kramskoy, Ivan 42
Krasznahorkai, László
   Satantango 4, 79–90
Krzysztofik, Robert 115
Kundera, Milan 10
Kupferberg *see* Miedzianka
Kurapaty *see* Kuropaty
Kurenevka 166
Kuropaty 163, 179
Kursk 128
Kühne, Olaf 170
Kyshtym *see* Chelyabinsk

Lane, Peter 37, 96
Lanzmann, Claude 162, 169, 172, 174, 180
'last foresters' *see* cursed soldiers
Latour, Bruno 134
Law and Justice Party (Poland) 196, 198
Lebda, Magdalena
   The Forest's Border 189
Lee Klein, Kerwin 27
Lenin, Vladimir 20, 41, 192
Leonov, Aleksei 73
Leszczyński, Witold 74
Lindbladh, Johanna 29–30, 148–9
Lithuania 161, 167, 193, 198–200, 203, 206
long-take cinematography *see* slow cinema
longue durée see deep-time, geology, natural or geological history
Łuskino, Leon
   Szara piechota (Grey Infantry) 168

McKenzie, Wark 44–5, 51, 54–5
McLellan, B.N. 194
McNeill, J. R. 2, 19, 21–3, 40, 96, 100, 103, 128, 131, 134
Magnitogorsk 45, 94
Makarski, Antoni
   About the Soldier 168
Mali, Barbara 115
Małczyński, Jacek 184–6
Mannan, Sam 128
Manser, Roger 10
Masing-Delic, Irene 62, 64, 66
Maslowski, Michel 10
Massumi, Brian 163
Mauldin, Erin Stuart 21–2
Mayak *see* Chelyabinsk
meadows 10, 71–2, 153, 167–9, 174, 178, 187, 204
meat 35–6, 51–3, 60, 66, 70–1, 109, 131, 198
Mednoye 179
Medvedev, Zhores 128–31, 142–4
memory
   anthropocentric discourse 169, 171, 185–7
   carrier 26, 28–9, 31
   collective 27
   earth 98, 104–5, 117
   forgetting 28, 81, 166–7
   gaps 25, 103, 115, 117–18, 120–1, 133, 174
   greening 186, 188
   landscape of memory 5, 175, 201, 205
   non-sites 162, 164, 172, 174–5, 180, 184
   places 10, 163, 166, 170, 174–5, 178, 186
   poetics 18, 30
   politics 167
   remembering 26–8, 166–7
   sites 26, 163, 170–2, 174–5, 181, 184–8
   studies 25–8, 31, 162–4, 167, 170–1, 174, 188
   see also disturbed landscapes, environmental memory
Michalak, Anna 98, 123
Mickiewicz, Adam
   Pan Tadeusz or the Last Foray in Lithuania 200–3
   see also Białowieża Forest

Miedzianka 103, 115–23
  see also copper, mining, Silesia, uranium
Miłosz, Czesław
  Odbicia (Reflections) 173–4
mining
  ecocriticism 99, 101–2, 113, 124
  etymology 98
  culture 97, 113–14
  history 93–7, 99, 115, 123
  landscape 100
  literature 97, 101–2
  mine 93, 95–7, 102, 108, 110–12, 118–21
  miners 42, 97, 104, 110–11, 116, 119–20
  narrative 94, 97–102, 123
  see also coal, geology, Miedzianka, Silesia, uranium
Młyńczak, Halina 181
Mońko, Michał 120
Moon, David 19–20
Moore, Irina 167
Morcinek, Gustaw 99
  Łysek z Pokładu Idy (Lysek from the Ida Shaft) 102
Morton, Timothy 2, 35, 102, 114, 134–5, 154
  see also hyperobjects
Moscow 40–1, 76
Motyka, Grzegorz 178
mountains 4, 10, 75, 94, 98, 100, 103, 115–18, 120–3, 187
  see also Magnitogorsk, Miedzianka
Mozhaev, Boris
  Lively 73
mud 74–6, 79–82, 85, 113
Murphy, Nick
  Chernobyl Nuclear – 'Surviving Disaster' 128
Muzaini, Hamzah 163
Mycio, Mary 132, 190
Myśliwski, Wiesław 71

Naimark, Norman 37
Narraway, Guinevre 86
nationalism 28, 80, 93, 162, 167, 178
  see also Białowieża Forest, coal, Great Hungarian Plain, Lithuania, mining, Poland, *puszcza*, Silesia

nature
  anthropomorphization 12, 164, 169, 171–2, 175, 181
  grief 180
  monument (*also* memorial) 20, 169, 183, 202
  mourning with 5, 162, 168, 177, 180, 182, 187–8
    see also Katyń Forest
  obstacle for commemoration 164, 166–7, 169, 171, 173, 175
  romanticized 35, 77–8, 162, 177, 183, 204, 206–7
    see also pastoral, *puszcza*
  shelter *see* forest, *puszcza*
  transformation 50, 62–4, 69, 71, 74
  vitality (*also* vegetation) 132, 169, 171, 173–4
  war on 45, 50
  witnessing 3, 5, 10, 16, 51, 163
Neirick, Miriam 12
Nekrich, Aleksandr 36, 39, 53–4
Nemes, László
  Son of Saul 184
Newerly, Igor 203
Niklasson, Mats 190
Nikulin, Dmitri 26–8
Nixon, Rob 127, 130–1, 137, 151–2
  see also slow violence
Niżyńska, Joanna 10
non-anthropocentric perspective (*also* ecocentric) 17, 23, 69, 97, 163, 184–5, 188, 206
  see also ecocritical perspective
non-human
  actors (*also* agents) 11, 17, 23–4, 55, 113–14, 123, 127
  death 173
  narrative (*also* perspective) 17, 88, 123
  subjects 22–3
  testimony 55, 103, 123
    see also environmental memory
  victim 50, 136
  voice (*also* narrator) 23, 30, 39, 90, 97, 107–8, 110, 112–13, 115
    see also polyphony
  witness 10, 51, 90, 123
  world 9, 20, 49, 52, 54, 88, 131, 147, 173
    see also ecological world

Nora, Pierre 26–7, 164
nuclear catastrophe *see* Chernobyl
   cities 127
   radiation *see* Chernobyl
   risk narrative *see* Chernobyl
   waste *see* Chernobyl's decontamination

Okhrana prirídy (a journal) 20
Orthodox Church 50, 63, 155
Orwell, George 58
   Animal Farm 63
Ostashevsky, Eugene 62
Oświęcim 184
   see also Auschwitz-Birkenau

Pakhmutova, Aleksandra 206
Parthé, Kathleen F. 59, 68, 73
pastoral
   myth 203, 205–6
   nature 162, 177, 183–4, 188, 201
   peasants 76–8
   tradition 192
   village 57, 62, 67–8, 70, 73–4
   see also Białowieża Forest, landscape, nature, peasantry, *puszcza*, rural culture
Patey-Grabowska, Alicja
   Impresje Katyńskie (Katyń Impressions) 180
Pavlov, Alexei Petrovich 101
   see also Anthropocene
peaceful atom 100
peasantry
   backward 74–5
   beliefs 74–5
   degeneration 87
      see also Satantango
   environmental culture 57
   kulaks 39, 60–1, 63
   pre-Soviet history 35–6
   resistance to collectivization 37, 49, 51–3, 60, 73–4, 78
   romanticized 49, 68, 76
   in Russia 36, 41, 67
   slaves 64, 81–2
   see also collectivization, rural culture
Pesniary (a musical band) 206–7
Pick, Anat 86
Pilnyak, Boris 36

Plan for the Transformation of Nature 9, 40, 62, 189
   see also collectivization, nature transformation, Stalin
Platonov, Andrey 43–5, 54–5
   Chevengur 44
   magical realism 56
   The Foundation Pit 4, 39, 42, 45–7
   The Wanderer 46
   Vprok – Bednyatskaya khoronika 43
   see also collectivization, Stalinocene
Plumwood, Val 177
Podlasye 162
Poland 190, 193, 197, 199
   see also Białowieża Forest, Katyń Forest
Polesie State Radioecological Reserve 153
Polessya (or *Polesye, Polesya*) 151, 178
Polish Central Statistical Office 93
Polish–Lithuanian Commonwealth 199–200, 206
polyphony 4, 37, 136, 149, 207
   see also nature, non-human voice
Pollack, Martin
   Kontaminierte Landschaften 1, 164, 169, 171–2, 174, 185
Polonization 99, 108
   see also Regained Territories, Silesia
Połońska, Elżbieta
   In the Noose 45
Pomian, Krzysztof 15–16
Poolos, Jamie 104
post-human *see* non-anthropocentric perspective, non-human
Pripyat 128, 131, 157
   see also Chernobyl
Prokofiev, Alexander 158
proletariat 42–3, 48, 55, 95
propaganda 20, 44, 54–5, 66, 74, 103, 128, 139, 158, 188, 204
   newspeak 52, 58, 60–2, 79
   newsreels 4, 40–1, 58, 60–1
   production novel 42
   see also contaminated language, Silesia
Pryde, Philip Rust 10
puszcza 190–3, 196, 198–206
   see also Białowieża Forest, pastoral
puszta see Great Hungarian Plain
Pyssa, Justyna 103

Raffnsøe, Sverre 101
rain 61, 79, 81, 83–5, 87, 121–2, 131, 142, 144, 151
   rainscape 80
   see also Chernobyl
Rapson, Jessica 177
Rasputin, Valentin 73
Red Army 64, 103
Redliński, Edward
   Konopielka 4, 74–6
Regained Territories 99, 199
   see also Polonization, propaganda, Silesia
René, Miloš 103
rewilding 131–3, 194
   see also Chernobyl
Rigby, Kate 134, 140
Rilke, Reiner Maria 72
Ritchie, Hannah 36
rivers
   Black River 30
   Bug 161
   flood 75, 184
   Narew 74–6, 161
   Odra 161
   San 161
   Soła 184
   Vistula 161, 185
   Volga 128
Rocky Mountain Arsenal Wildlife Reserve 131–2
Rodak, Anna
   Las (Forest) 182–3
Roser, Max 36
Roskin, Michael G. 10
Rothberg, Michael 26, 170
Royer, Clara 184
Różewicz, Tadeusz 99
Różycki, Tomasz
   Scorched Maps 173–4
Rueckert, George 61, 69
rural culture 4, 36–7, 66–9, 73–8, 151
   cultural memory 57, 68, 76
   ecocriticism 57, 62, 69, 74
   literature 57, 59, 71, 73, 77, 90
   soil 36–7, 41, 45, 51, 57, 67–8, 73
   see also collectivization, pastoral, peasantry
Russian cosmism 64, 66, 70, 72

Russian soul 41–2, 95, 150
Rymkiewicz, Jarosław Marek
   Polish Conversations in Summer 1983 199

Samizdat (underground press) 199

Samojlik, Tomasz 190, 195
Schama, Simon
   Landscape and Memory 191–2, 199, 201, 203, 206
Schramm, Katharina 164
Second World War 3, 9, 96, 99, 102, 104, 109, 115–18, 120, 122, 163
Seifrid, Thomas 43–7, 50
Sendyka, Roma 169, 171–2, 174–5, 180
Shalamov, Varlam
   Kolyma Tales 97
Shklovsky, Viktor 44, 95
shock workers (a Stakhanovite or normbuster) 58–9
   see also Stakhanov, Aleksei
Sholokhov, Mikhail 44
Siberia 20, 100, 187
Silesia
   cultural memory 94, 101–2, 104, 107, 112, 114, 123
   environmental history 97–8, 101–2, 107, 113–14, 122–3
   identity 107, 111–13
   literature 99, 101–2, 107, 123
   Lower S. 4, 94, 103, 115
   mining culture 108, 113–14
   mythology (*also* folk) 112–13, 115
   Upper S. 94, 99, 102, 107
   see also coal, mining, Polonization, Regained Territories, uranium
silver 103, 116
Sirota, Lyubov 129
Six-Year Plan 97
Skłodowska-Curie, Maria 103
Śliwiński, Tomasz 185
Slovakia 96, 194
slow cinema 85–8
slow violence 130–1, 135, 137, 151
Słoński, Edward
   On the Ashes 168–9
Smolensk 179
Smykowski, Mikołaj 172

Snyder, Timothy
  Bloodlands 1, 165, 169, 185
Sobibór 170
socialism 9, 19, 21, 30, 43–50, 63, 102, 104, 198
  see also collectivization, communism
socialist economy 35, 41, 58, 96–7
Socialist Realism 42, 59–60, 62–3, 75
soil 11–12, 59, 93, 108–9, 131
  see also collectivization, peasantry, rural culture
Soloukhin, Vladimir
  To Make Birds Sing 73–4
Solzhenitsyn, Aleksandr
  Matryona's Home 73
  The Gulag Archipelago 97
Soroczyński, Tadeusz 204
Soviet Bloc (*also* Eastern Bloc) 9–10, 20, 25, 37, 40, 42, 93–4, 96, 101, 118, 136, 199
sovietization 37, 74, 163, 202
Soviet Eastern Europe
  cultural memory 10, 31, 206
  energy crisis 94, 98
  environmental cultures 13, 24, 90
  environmental history 5, 10, 23
  literature 59
  'unknownland' 9, 12
Soviet Union 9, 40, 93, 96, 108, 194
  collapse 10, 127, 191
  crisis 79, 86, 89, 129, 131, 158
  stigmatization 10, 13, 19–21
Sörlin, Sverker 22
Springer, Filip 104
  History of a Disappearance. The Story of a Forgotten Polish Town 103, 115–24
  see also Kupferberg, Miedzianka, uranium
Stakhanov, Aleksei 42, 58
  Stakhanovite record 42
Stalin, Joseph Vissarionovich 3, 20–1, 47, 50, 67, 73, 97, 103–4, 161
  death 40, 59, 61
  environmental politics 9, 41, 45, 189–90, 192
    see also Plan for the Transformation of Nature
  name 40–1

  purges 36–7, 39, 59, 179, 183, 187
Stalinism 9, 21, 42, 50, 54, 60–2, 90, 149
Stalinocene 19, 39–47, 54
  see also Five-Year Plan, hyperindustrialization, industrialization
Stasiuk, Andrzej 10
  The East 76, 185
steppe 43–4, 60, 80, 95, 189
Stoetzer, Bettina 162, 164
Sturdy, Caroline 186
Szczepan, Aleksandra 172
Szpak, Ewelina 162
Szwagrzyk, Jerzy 195

Tarkovsky, Andrei
  Stalker 153
Tarkovsky, Arseny 69
Tarr, Béla
  Satantango 4, 79, 81–2, 85–9
Teksty Drugie (a journal) 187
Thaw 59, 73
tired village 4, 37, 39, 79, 90
  see also collectivization
Tokarczuk, Olga
  Drive Your Plow Over the Bones of the Dead 197
Tomiałojć, Ludwik 194
Topolski, Jerzy 15
totalitarianism 50, 55, 203
  see also Stalinism
Traba, Robert 10, 170
trees 25, 131, 158, 174, 178, 202
  judges 165–6, 169
  Katyń poetry 180–3, 186
  national monuments 195
  in Polish patriotic songs 168
  see also Białowieża Forest, Chernobyl, forest, Katyń Forest, *puszcza*
Trexler, Adam 114, 134
Trojanowska, Tamara 10
Tsiolkovsky, Konstantin 64
Tsvirko, Lidia 153
Tuan, Yi-Fu 80, 174–5
Tumarkin, Maria 163
Turgenev, Ivan 42
Turnock, David 10
Twardoch, Szczepan
  Drach 4, 102, 104, 107–14

Ubertowska, Aleksandra 184
Ukraine 36–7, 128, 144, 173, 178–9
Ukrainian Insurgent Army (UPA) 178
UNESCO World Heritage Site 195–6
Unger, Corinna 19
uranium
  extraction 93, 96, 117–18
  health damage 103, 119–20
  history of 'Polish uranium' 102–4, 115, 117
  narrative 94, 98, 115, 118
  patchy memory 120–1
  see also Kletno, Kowary, Miedzianka, mining, Silesia
USSR *see* Soviet Union
utopia
  collectivization 37, 42, 64, 66
  idealism 44–5, 48, 50, 158
  optimism 95, 143
  'the land-utopia' 9
  vision of nature 20, 54, 62–4

van Alphen, E. 171
Veyne, Paul 11–12
Village Prose 57, 73–4, 79
Viskuli 192
  see also Belavezha Accords
Vitale, Serena 95
Volhynia 5, 162, 178, 188
  see also disturbed landscape
Volhynia Slaughter 178–9
Voznesensky, Andrei 95
  After the Tone 29–30

Wallace, Molly 127, 132, 134
Wantuła, Leon
  Urodzeni w dymach (Born in Smoke) 99
Warburg, Aby 26
Warde, Paul 22
Warsaw 162, 173
Warsaw Pact 10
Wägenbaur, Thomas 30
weather 11, 79, 81, 83, 86, 144

Webster, Barbara Jancar 10
Weijde, Erik van der
  Der Baum 174
Weiner, Douglas 20–1, 40
Wesołowski, Tomasz 194
wetlands 4, 44, 74–5, 131
White, Hayden 16–17, 27
White Sea Canal 40, 45
Wigry Lake 199
Wilke, Sabine 40
wind 49, 56, 79, 81, 83, 85–6, 151, 181
Wirth, Peter 115
Wittlin, Józef
  The Salt of the Earth 111
Wolf, Christa
  Accident: A Day's News 4–5, 136–45
Wołk-Jezierska, Witomiła
  Garstka ziemi (A Handful of Soil) 180

Yeoh, Brenda 163
Yevtushenko, Yevgeny 55
  Babiy Yar 164–6
Young, Georges M. 64
Yuzefpolskaya, Sofija 61, 69

Zabolotsky, Nikolai 57, 61, 69, 72, 89
  A Walk 69–70, 72
  Agriculture Triumphant (or *The Triumph of Agriculture*) 4, 61–6, 69
  I Do not Look for Harmony in Nature 69
  Metamorphoses 187–8
Zabuzhko, Oksana 36, 136, 144, 147
Zapf, Hubert
  culturecology 30–1
zapovednik (nature reserve) 20, 153, 189–90
Ziegler, Charles 10
Zink, Andrea 130, 148–9, 156
Zub, Karol 194
Zvezda (journal) 61
Żeromski, Stefan 203
Żywiec (city) 184

www.ingramcontent.com/pod-product-compliance
Lightning Source LLC
Chambersburg PA
CBHW072146290426
44111CB00012B/1987